GERMAN TANKS
OF WORLD WAR TWO

CRAIG MOORE AND DAVID BOCQUELET

MILITARY VEHICLES AND ARTILLERY SERIES, VOLUME 1

Published by Key Books
An imprint of Key Publishing Ltd
PO Box 100
Stamford
Lincs PE19 1XQ

www.keypublishing.com

The right of Craig Moore to be identified as the author of this book has been asserted in accordance with the Copyright, Designs and Patents Act 1988 Sections 77 and 78.

Copyright © Craig Moore, 2020

ISBN 978 1 913295 74 5

20 21 22 23 24   10 9 8 7 6 5 4 3 2 1

All rights reserved. Reproduction in whole or in part in any form whatsoever or by any means is strictly prohibited without the prior permission of the Publisher.

Typeset by SJmagic DESIGN SERVICES, India.

Acknowledgements
Herbert Ackermans, Peter Chamberlain, Rob Cogan, Hilary Louis Doyle, Marcus Hock, Thomas L. Jentz, Walter J. Spielberger, Bovington Tank Museum, Imperial War Museum, U.S. Army Armor & Cavalry Collection at Fort Benning.

Unless otherwise stated, all images are the copyright of David Bocquelet.

Contents

Introduction		4
Chapter 1	Neubaufahrzeug (Neubau-Panzerkampfwagen IV)	8
Chapter 2	Panzer I Ausf.A and Ausf.B (Sd.Kfz.101)	10
Chapter 3	Panzer I Ausf.C, Ausf.F and Panzerbefehlswagen (Sd.Kfz.101)	12
Chapter 4	Panzer II Ausf.a (Sd.Kfz.121)	14
Chapter 5	Panzer II Ausf.b (Sd.Kfz.121)	16
Chapter 6	Panzer II Ausf.c (Sd.Kfz.121)	18
Chapter 7	Panzer II Ausf.A (Sd.Kfz.121)	20
Chapter 8	Panzer II Ausf.B (Sd.Kfz.121)	22
Chapter 9	Panzer II Ausf.C (Sd.Kfz.121)	24
Chapter 10	Panzer II Ausf.D (Sd.Kfz.121) and Ausf.E	26
Chapter 11	Panzer II Ausf.F (Sd.Kfz.121)	28
Chapter 12	Panzer II Ausf.G (Sd.Kfz.121/1)	30
Chapter 13	Panzer II Ausf.J (Sd.Kfz.121)	32
Chapter 14	Panzerspähwagen II (2cm) 'Luchs' (Sd.Kfz.123)	34
Chapter 15	Panzer 35(t)	36
Chapter 16	Panzer 38(t)	38
Chapter 17	Panzer III Ausf.A–D (Sd.Kfz.141)	40
Chapter 18	Panzer III Ausf.E (Sd.Kfz.141)	42
Chapter 19	Panzer III Ausf.F (Sd.Kfz.141)	44
Chapter 20	Panzer III Ausf.G (Sd.Kfz.141)	46
Chapter 21	Panzer III Ausf.H (Sd.Kfz.141)	48
Chapter 22	Panzer III Ausf.J (Sd.Kfz.141) and Ausf.L (Sd.Kfz.141/1)	50
Chapter 23	Panzer III Ausf.M (Sd.Kfz.141/1) and Ausf.N (Sd.Kfz.141/2)	52
Chapter 24	Panzer IV Ausf.A (Vs.Kfz.622)	54
Chapter 25	Panzer IV Ausf.B (Vs.Kfz.622)	56
Chapter 26	Panzer IV Ausf.C (Vs.Kfz.622)	58
Chapter 27	Panzer IV Ausf.D (Sd.Kfz.161)	60
Chapter 28	Panzer IV Ausf.E (Sd.Kfz.161)	62
Chapter 29	Panzer IV Ausf.F (Sd.Kfz.161) and F2 (Sd.Kfz.161/1)	64
Chapter 30	Panzer IV Ausf.G (Sd.Kfz.161/1)	66
Chapter 31	Panzer IV Ausf.H (Sd.Kfz.161/2)	68
Chapter 32	Panzer IV Ausf.J (Sd.Kfz.161/2)	70
Chapter 33	Panzer IV mit Hydrostatischem Antrieb	72
Chapter 34	Panther Ausf.D (Sd.Kfz.171)	74
Chapter 35	Panther Ausf.A (Sd.Kfz.171)	76
Chapter 36	Panther Ausf.G (Sd.Kfz.171)	78
Chapter 37	Tiger I (Sd.Kfz.181)	80
Chapter 38	Tiger II (Sd.Kfz.182)	88
Chapter 39	Panzer VIII Maus	94
Conclusion		96

Introduction

During World War Two, the British Army provided their troops with a guidebook on armoured fighting vehicle (AFV) recognition (see pages 5, 6 & 7). It included information on enemy tanks and other military vehicles. These books also included data on those used by Allied forces, to help reduce the amount of so-called 'friendly fire' incidents, when troops fired on their own side's vehicles that were wrongly identified as belonging to the enemy.

New editions were printed regularly throughout the war as more intelligence information became available. Photographs, sketch drawings and handwritten notes covering recently captured or knocked-out new German tanks were quickly added to the guidebook. When reading these books, it is noticeable that there were gaps in the British Army's knowledge of German tanks during specific periods of World War Two.

There were lots of changes made to tanks throughout their production life. This is why in this book there are different pages on different versions of the same tank. The German word for model or version is 'Ausführung'. This is often abbreviated to 'Ausf.'.

Many of the changes made to German tanks were as a result of feedback from the tank crews on the battlefield. For example, the smoke grenade launchers on the side of tank turrets were withdrawn after a Tiger tank crew reported that their smoke grenades exploded in their launchers after being hit by Soviet small-arms fire. The tank was surrounded by smoke so the crew could not see where the enemy were and nearly suffocated inside the tank. More powerful anti-tank guns and thicker armour were added to German tanks as the threat they had to deal with increased. The arrival of the Soviet T-34 and KV tanks resulted in many German tanks being upgraded or no longer used as they were now out-classed and considered obsolete. Some of the changes were made to speed up the production of tanks in the factory. Others were made to cope with the lack of raw materials, such as rubber, or different types of metal.

Identifying a tank by its Ausführung letter (Ausf.A, Ausf.B, Ausf.C, Ausf.D, etc.) is helpful, but changes were made all the time on the production line. For example, you will often read that a tank was an early-production Ausf.G, early-middle-production Ausf.G, late-middle-production Ausf.G or a late-production Ausf.G. This is because that tank had different features added or taken away at four stages during the production run of the Ausf.G.

It is confusing, but to make things worse, you will see photographs of tanks produced in 1943 with fixtures and fittings that were not introduced until 1944 or 1945. Sometimes, tanks made in 1945 will have parts salvaged from damaged tanks built in 1943 or 1944. When a tank was damaged, it would be fixed with the spare parts that were available in the workshop stores. Many early tanks were backfitted with new equipment, additional armour and stowage bins on turrets after they had left the factory.

The aim of this book is to give the reader a better understanding of German tanks that were used in combat and the modifications that were made during their production run. This knowledge will assist them recognise different tanks. Sturmgeschütz assault guns, Jagdpanzer hunting tanks, self-propelled anti-tank guns, self-propelled artillery guns, Beutepanzers captured tanks in German service and prototype tank designs are not include in this volume.

The images in this book are an artist's illustration and not a technical scale drawing.

Introduction

AFV Recognition
Part II. July, 1943
Enemy turreted AFVs

RESTRICTED

The information given in this document is not to be communicated, either directly or indirectly, to the Press or to any person not authorized to receive it.

Prepared under the direction of The Chief of the Imperial General Staff
THE WAR OFFICE   1943

CROWN COPYRIGHT RESERVED

26/G.S. Publications/932

AFV Recognition
Part III
Air/ground recognition of Allied and Enemy AFVs
November, 1944

RESTRICTED

The information given in this document is not to be communicated, either directly or indirectly, to the Press or to any person not authorized to receive it

Prepared under the direction of The Chief of the Imperial General Staff
THE WAR OFFICE   - - -   1944

Crown copyright reserved

GS/26 Publications/1308   21278

GERMAN TANKS

German Medium tank   Pz.Kw. III

POINTS FOR RECOGNITION
Broadside   *Head-on*

Turret
Squat with top tapering to rounded front. Ringed cupola on turret rear. Double doors in sides of turret. Undercut front corners. Sloping back with stowage bin. Possible spaced armour in front.

Squat and streamlined; ringed cupola centrally placed. Undercut front corners. Broad rectangular front.

Armament
Long-barrelled 5 cm gun mounted in rounded mantlet. Hull MG mounted on right-hand side of driver's compartment.

Long-barrelled 5 cm gun and co-axial MG on its right mounted in broad rectangular mantlet. Hull MG mounted on right-hand side of driver's compartment.

Track assembly
Six small bogie wheels of equal size and evenly spaced. Very large front sprocket. No suspension springs visible. Small door in side hull is dispensed with in latest models.

Hull
Low superstructure. Nearly vertical front to driver's and front gunner's compartment. Sloping rear deck. Spaced armour on front of superstructure.

Angular nose-plate. Low superstructure. Clean rectangular front to driver's and front gunner's compartment with spaced armour in all recent models.

Remarks.—Earlier types have a shorter barrelled 5 cm gun and very early models have eight bogie wheels. The various models differ little from the recognition point of view; the differences being mostly confined to (a) spaced armour, (b) designs of sprocket and idler wheels, (c) design of hull MG mounting, (d) driver's visor. The chassis is widely used to mount SP artillery guns. Recent models frequently mount a short-barrelled (telescope shaped) 7·5 cm gun.

GERMAN TANKS

German Medium tank Pz.Kw. III

Armour, thickest plate 70 mm (of which 20 mm is spaced armour)
Weight, laden ... 22 tons
Crew 5
Armament, turret 5 cm or short 7·5 cm gun, 1 MG
Armament, hull ... One MG
Overall length (without gun) 17 ft 9 in
Overall width ... 9 ft 8 in
Overall height ... 8 ft 3 in
Engine ... ... 300 bhp petrol
Speed, road ... 25 mph
Radius of action, road 100 miles
Trench crossing ... 8 ft 6 in
Step ... ... 2 ft 0 in
Fording ... ... 2 ft 8 in

With short 7·5 cm gun, mantlet variation, and smoke dischargers.

German Pz.Kw.III (Models E, F, G) (Similar layout on later types)
Wireless mast
"Hull" MG
Driver's visor
Gun mantlet shield
Superstructure front
Side hull plate
Upper and lower nose plates
Front belly plate
Glacis plate
1 3¼"
9'7"
Five vision openings
Bin
Pistol port (each side)
Air intake
8'3"
Rear & main belly plates
Main belly plate
0 1 2 3 4 5 6 7 8 Scale in feet
17'9"
Upper nose plate
Superstructure roof
Rear deck
Tail plate
Five smoke candles
Air INTAKE louvres (good bomb aiming points)
ARMOUR:*
Heavy
Medium
Light
* as fitted on latest models – Key to diagrams only
Tail plate

German Pz.Kw. VI "Tiger" Heavy tank
VARIATIONS
SHOWING HOW and WHY two widths of this AFV may be encountered
Normal or wide tracks may be used.
Gun shown without muzzle brake.
Hull without sand-shields
ORIGINAL TRACK
HULL
SECTION OF TRACK
The addition of detachable wheels for BROAD TRACK also gives extra protection below
IDLER
SPROCKET
DETACHABLE WHEELS
sand-shields
Pz Kw II
III
IV
VI
DEVELOPMENT of GERMAN AFVs
Note II III & IV have same size sprockets.
LARGE SIZE FRONT SPROCKETS are a feature of German tanks.

Progressive types of the German Pz.Kw.III Medium tank
Early model
(3.7cm gun and note recessed mantlet shield)
dummy gun
– as used now for some commanders' models
Later models
had leather jacket on 5cm gun and round perforations in sprocket.
This model differs in superstructure front, sprocket and idler. Jockey wheels not equidistant
The addition of spaced armour bolted to mantlet shield and superstructure front almost coincided with advent of longer 5cm gun.
The actual extra shield has often not been fitted to its frame.

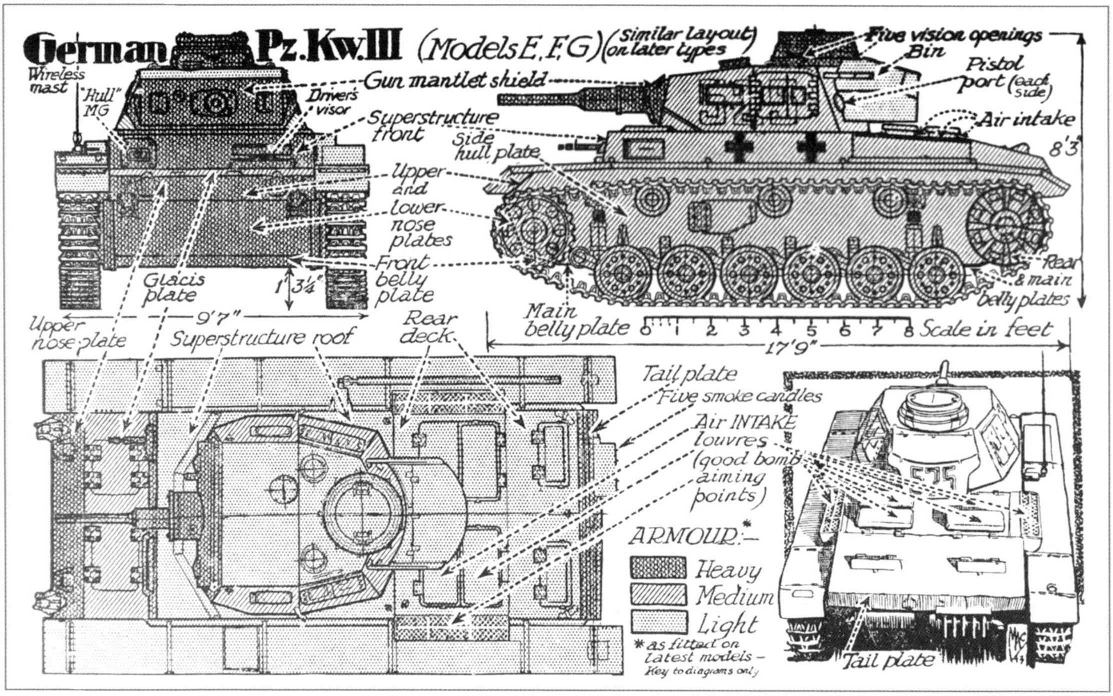

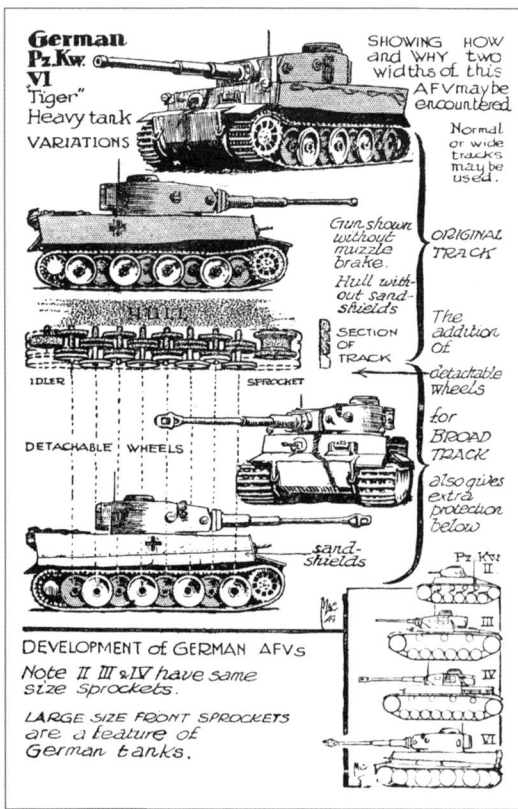

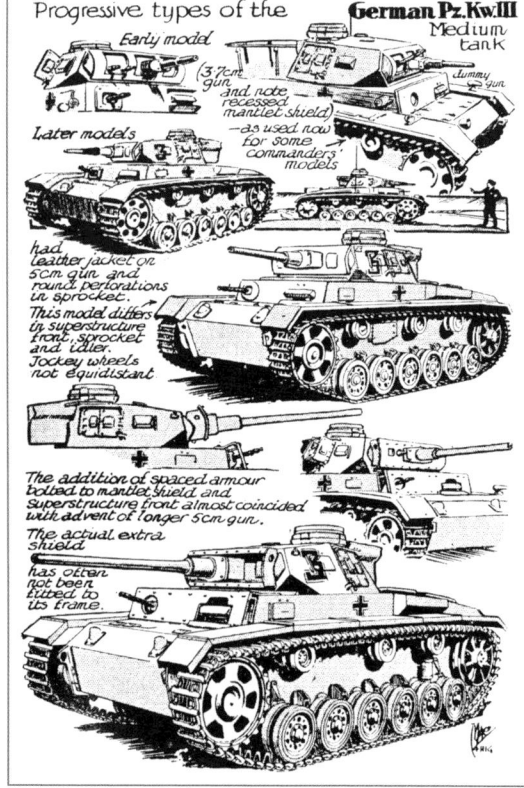

Introduction

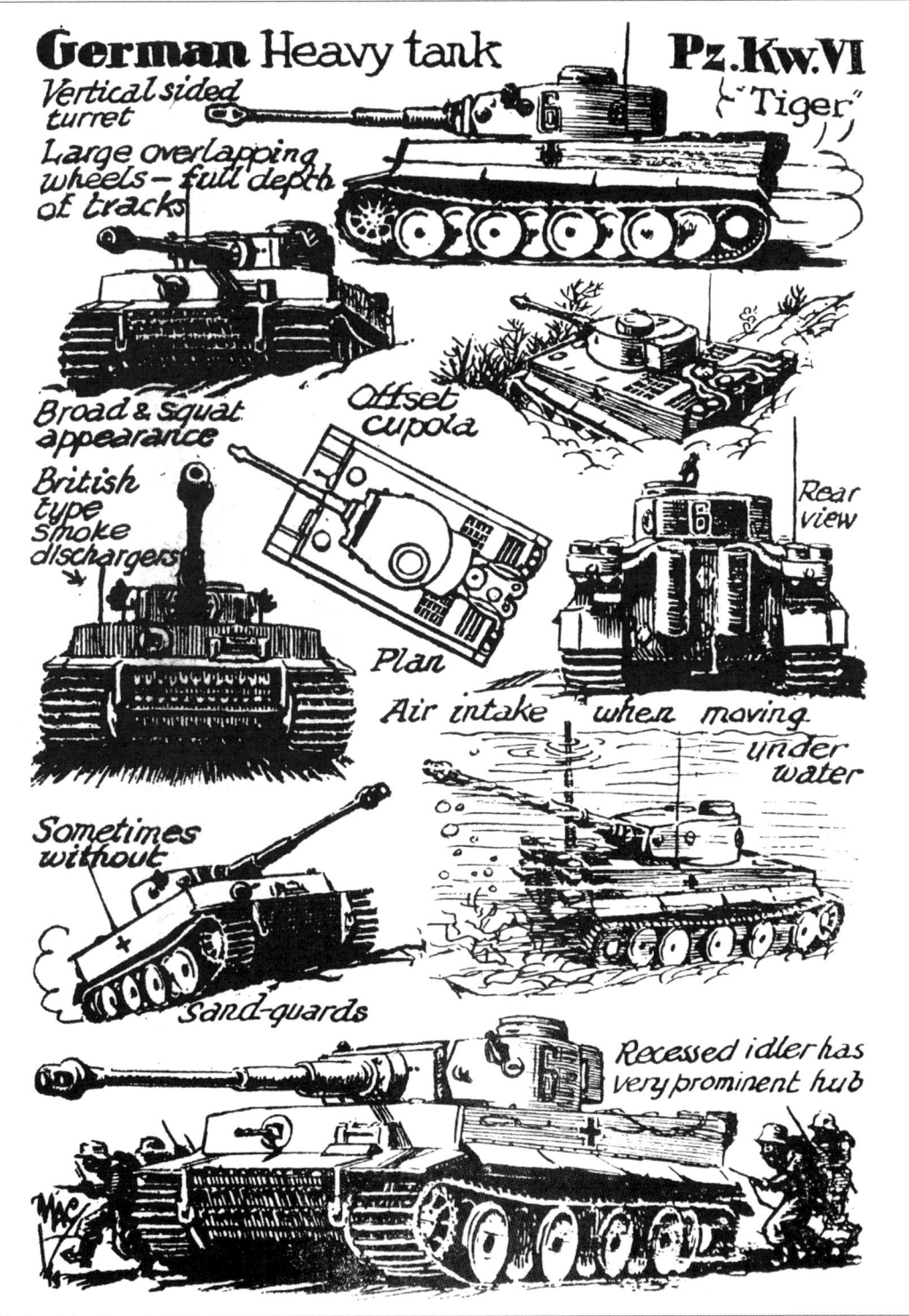

These illustrations were commissioned for inclusion in the British War Office informative booklet called *AFV Recognition Part II. July, 1943 Enemy turreted AFVs*.

Chapter 1
Neubaufahrzeug (Neubau-Panzerkampfwagen IV)

When Adolf Hitler came to power, for propaganda purposes he needed symbols of Germany's renewal as a powerful nation. He used the new Neubaufahrzeug tank as one of those symbols. The name 'Neubaufahrzeug' literally translates to 'new construction vehicle', and 'Panzerkampfwagen' means 'armoured combat vehicle'.

After Germany's defeat in World War One, the terms of the Versailles Treaty prevented the country rearming. To hide the fact that they were developing a tank, it was called a Großtraktor (large tractor). Between 1929 and 1935, four prototypes were trialled, first in the Soviet Union and later in Germany. Mechanical defects plagued their deployment and they were turned into impressive-looking statues after the trials were finished.

New prototype designs were ordered. Rheinmetall manufactured the Neubaufahrzeug Nr.1 hull using soft steel and fitted it with a Rheinmetall-designed round turret. It had a 3.7cm gun mounted over the top of the 7.5cm Kw.K. L/24 gun in the turret. The Neubaufahrzeug Nr.2 hull was also built by Rheinmetall using soft steel, but it was fitted with an angular turret designed by Krupp. It also had a 7.5cm Kw.K. L/24 main gun in the turret with a coaxial mounted 3.7cm Kw.K. L/45 gun in a single mantlet. Two 7.92mm M.G.34 machine guns were mounted in two smaller turrets, one on the right of the driver's position and the other behind the main turret to cover the rear of the tank. A third was mounted in the main turret to the right of the 7.5cm Kw.K. L/24 gun. The Neubaufahrzeug with the Krupp turret was the preferred design. Rheinmetall Neubaufahrzeug Nr.3, 4 and 5 with a Krupp turret were built with armour plate and renamed Neubau-Panzerkampfwagen IV. After further tests, the design was rejected. It was decided to concentrate on producing the more agile Panzer III and Panzer IV medium tanks instead of this slow, heavy tank.

On 19 April 1940, the three Neubau-Panzerkampfwagen IV tanks were sent to Norway. They looked impressive and were photographed for propaganda. They took part in local security operations and attacked British positions near Kvam on 25 April 1940. One tank was disabled by an anti-tank mine and scrapped in situ. The two remaining tanks were captured in 1945 and later scrapped.

Specifications

Length:	6.65m
Width:	2.90m
Height:	2.90m
Weight:	23 tonnes
Engine:	BMW Va 6-cylinder 290hp petrol/gasoline
Crew:	6
Main gun:	7.5cm Kw.K. L/24
2nd gun:	3.7cm Kw.K. L/45
Other weapons:	3 x 7.92mm M.G.34 machine guns
Armour:	8mm–20mm
Max. road speed:	30km/h
Max. range on roads:	120km
Total built:	3 (+ 2 soft steel)

Neubaufahrzeug (Neubau-Panzerkampfwagen IV)

Rheinmetall Grosstraktor (*Großtraktor*) prototype.

Rheinmetall Neubaufahrzeug Nr.1 hull with Rheinmetall turret (soft steel).

Rheinmetall Neubaufahrzeug Nr.2 hull with Krupp turret (soft steel).

Neubau-Panzerkampfwagen IV. Three Rheinmetall Neubaufahrzeug hulls Nr.3–5 with Krupp turrets were built with armour plate and renamed Neubau-Panzerkampfwagen IV.

Chapter 2
Panzer I Ausf.A and Ausf.B (Sd.Kfz.101)

Germany secretly started to rearm from the mid-1920s. The Panzer I was the first mass-produced tank of the German Army. Initially, it was intended to be used as a driver training tank: the first 15 were delivered without turrets. Its full name was Panzerkampfwagen I Ausführung A, Sonderkraftfahrzeug 101 (armoured combat vehicle I version A, special purpose vehicle 101). This was abbreviated to PzKpfw I Ausf.A (Sd.Kfz.101).

The turreted version of the tank was armed with two 7.92mm MG13K machine guns. Production started in 1934. The tank's armour ranged from 5mm to 15mm thick, which would only protect the crew from small-arms fire. It was used as a reconnaissance scout tank.

The next version of the tank was the Panzer I Ausf.B. It was 40cm longer to accommodate a more powerful 100hp engine and had an additional 5th road wheel. The armour thickness and weapons remained the same. It was slightly heavier: its combat weight was now 5.8 tonnes rather than 5.4 tonnes. It was fitted with a more powerful Maybach engine to cope with the weight increase.

Panzer I tanks first saw action during the Spanish Civil War in 1936 with the Condor Legion at the Battle of Madrid. A Panzer I could only damage the Soviet T-26 tanks at short range. Both versions of the tank were used in the 1939 invasion of Poland, the 1940 occupations of Denmark and Norway and the 1940 Battle of France.

In 1941, they were shipped to North Africa where they saw action in the desert with Rommel's Afrika Korps as scout tanks. They were later replaced by the Panzer II. They were also used on the Eastern Front but were withdrawn from front-line duties in the spring of 1942 after it was found they could not cope with the poor winter conditions. They continued to be used for tank crew training and internal security anti-partisan policing roles.

Specifications

	Ausf.A	Ausf.B
Length:	4.02m	4.42m
Width:	2.06m	2.06m
Height:	1.72m	1.72m
Weight:	5.4 tonnes	5.8 tonnes
Engine:	Krupp M 305 4-cyl air-cooled 3.5 litre 60hp petrol/gasoline	Maybach NL 38 Tr 6-cyl water-cooled 3.8 litre 100hp petrol/gasoline
Crew:	2	2
Left barrel:	7.92mm M.G.13K machine gun	7.92mm M.G.13K machine gun
Right barrel:	7.92mm M.G.13K machine gun	7.92mm M.G.13K machine gun
Armour:	5mm–15mm	5mm–15mm
Max. road speed:	37.5km/h	40km/h
Max. range on roads:	140km	170km
Total built:	1,190	399

Panzer I Ausf.A and Ausf.B (Sd.Kfz.101)

Panzer I Ausf.A.

Panzer I Ausf.A.

Panzer I Ausf.B.

Panzer I Ausf.B.

Chapter 3
Panzer I Ausf.C, Ausf.F and Panzerbefehlswagen (Sd.Kfz.101)

Early Panzer I tank hulls were converted into self-propelled anti-tank guns, artillery guns, anti-aircraft flakpanzers, ammunition and supply carriers, armoured ambulances and engineering tanks. They were also used as kleiner Panzerbefehlswagens (kl.Pz.Bef.Wg.) (Sd.Kfz.265) light command tanks. Around 190 of these PzBefw, fast command tanks, were built based on the Panzer I Ausf.B tank hull. They had a tall 1.99m armoured superstructure that housed a FuG6 radio transmitter and large frame antenna. They were armed with one 7.92mm M.G.34 machine gun.

Although still called the Panzer I, the Ausf.C version was a very different vehicle. It had torsion-bar suspension with large interleaved road wheels. It had a more powerful Maybach HL45 150hp engine. These new features gave the tank a fast top road speed of 79km/h even though the armour thickness had been doubled, compared to the Panzer I Ausf.B, to 30mm at the front of the tank.

A long-barrelled 7.92mm E.W.141 self-loading semi-automatic machine gun was mounted in the turret next to a standard 7.92mm M.G.34 machine gun. It was intended to be used by the Luftlandetruppen (Airborne troops) and the Kolonial Panzertruppen (Colonial Armoured Troops). In early 1943, two were sent to the Eastern Front for combat evaluation. In 1944, the other 38 were issued to LVIII Panzer Reserve Korps which fought in Normandy.

The Panzer I Ausf.F had additional protective armour and the front armour was now 80mm thick. It was intended to be used against fortified strongpoints and have a weight limit of 18 tonnes so that it could safely drive over army engineers' combat bridges, but it was 3 tonnes heavier. In September 1942, seven were reported as being used on the Eastern Front, near Leningrad with 1.Kp./Pz.Abt.z.b.V.66. Five more were sent in January 1943. Four were assigned to II.Abt.Stabskompanie on 9 May 1943 and five were issued to Pz.Inst.Abt.559 on 10 July 1943.

Specifications

	Ausf.C	Ausf.F	kl.Pz.Bef.Wg.
Length:	4.19m	4.38m	4.42m
Width:	1.92m	2.64m	2.06m
Height:	1.94m	2.05m	1.99m
Weight:	8 tonnes	21 tonnes	5.88 tonnes
Engine:	Maybach HL45P 150hp petrol/gasoline	Maybach HL45P 150hp petrol/gasoline	Maybach NL 38 Tr 100hp petrol/gasoline
Crew:	2	2	3
Left barrel:	7.92mm Einbauwaffe 141 MG	7.92mm M.G.34	
Right barrel:	7.92mm M.G.34	7.92mm M.G.34	7.92mm M.G.34
Armour:	5mm–30mm	20mm–80mm	8mm–14.5mm
Max. road speed:	79km/h	25km/h	40km/h
Max. range on roads:	300km	150km	170km
Total built:	40	30	184

Panzer I Ausf.C, Ausf.F and Panzerbefehlswagen (Sd.Kfz.101)

Panzer I Ausf.C.

Panzer I Ausf.C.

Panzer I Ausf.C.

Panzer I Ausf.F.

Chapter 4
Panzer II Ausf.a (Sd.Kfz.121)

In January 1934, the German tank design office of the weapons testing ordnance department Waffen Prüfwesen 6 (Wa Prüf.6) issued specifications of a new tank hull they wanted built, code name La.S.100. Weapons manufacturer Maschinenfabrik Augsburg Nürnberg AG (M.A.N.) constructed a prototype La.S.100 tank hull. They competed with two other German companies, Fried Krupp Abt.A.K. and Henschel. M.A.N. was awarded the contract to build the hull of the new Panzer II light tank based on their prototype La.S.100 hull. Daimler-Benz designed the superstructure and turret.

It is wrong to dismiss the Panzer II tank of 1936 as being poorly designed when comparing it with more heavily armed and armoured tanks of World War Two. The tank's armour could protect its crew from small-arms fire and 7.92mm S.m.K. steel-cored armour-piercing machine gun bullets fired from a range of 30m. It was designed to attack enemy machine-gun nests and destroy them so that the accompanying infantry could continue to advance. The Panzer II was not primarily designed to engage in tank on tank combat. The tank's 2cm Kw.K.30 L/55 gun could knock out Soviet T-26 and BT tanks but the crews were aware that the Panzer II tank's armour would not stop a 3.7cm or 4.5cm anti-tank gun.

The high nickel-alloy, rolled, homogeneous hard armour plate ranged in thickness from 5mm to 13mm. It was welded together not riveted as seen on many other tanks of this time-period. This made it stronger and lighter.

The first Panzer II Ausführung (model versions) were given the lower case letter 'a' then 'b' and 'c'. Later versions were given capital letters 'A', 'B' and 'C'. This can be confusing. The Panzer II Ausf.a tanks were subdivided into Ausf.a/1, Ausf.a/2 and Ausf.a/3. Each version had different minor mechanical changes.

Early versions of the Panzer II changed shape over time as they were upgraded during their operational life. Additional armour was added, and features such as cupolas were fitted. Panzer II tanks were not used in the Spanish Civil War. They first saw combat in Poland on 1 Sept 1939.

Specifications

Length:	4.38m
Width:	2.14m
Height:	1.94m
Weight:	7.6 tonnes
Crew:	3
Engine:	Maybach HL 57 TR 6-cylinder water-cooled 130hp petrol/gasoline
Armament:	2cm Kw.K.30 L/55 auto-cannon
Additional weapon:	7.92mm M.G.34 machine gun
Armour thickness:	5mm–15mm
Max. road speed:	40km/h
Max. range on roads:	190km
Total built Ausf.a/1:	25
Total built Ausf.a/2:	25
Total built Ausf.a/3:	25

Panzer II Ausf.a (Sd.Kfz.121)

Chapter 5
Panzer II Ausf.b (Sd.Kfz.121)

The thickness of armour on the Panzer II Ausf.b tank's hull, superstructure and turret were increased from the Ausf.a tank's 13mm to 14.5mm. The gun mantle increased from 15mm to 16mm. To reduce the reliance on being able to obtain nickel, the armour was changed to rolled, homogeneous, nickel-free steel armour. It had the same resistance to 7.92mm S.m.K. steel-cored armour-piercing machine gun bullets fired from a range of 30m as the Ausf.a, but it had to be thicker to achieve this. This increased the weight of the tank by 500kg, but it did not decrease its speed.

The shape and thickness of the crew's vision ports were changed to give added protection. A different style of large drive wheel was bolted onto the final drive at the front of the tank. The rear engine deck was redesigned. Armoured louvres were added to the rear right of the tank. The road wheels and the track-return rollers were widened. The return rollers were reduced in diameter. Wider tracks, increasing from 260mm to 285mm, were introduced. Lengthened, foldable track guards were fitted to the rear of the tank.

The 2cm Kw.K.30 gun could fire three different shells. When fired against armour plate laid back at 30° from the vertical, the PzGr.39 armour-piercing shell could penetrate 23mm of armour at 100m and 14mm of armour at 500m. The PzGr.40 armour-piercing composite rigid shell could go through 40mm of armour at 100m and 20mm of armour at 500m. It could also fire 2cm Sprgr.39 high-explosive shells.

Early versions of the Panzer II changed shape over time as they were upgraded during their operational life. Additional armour was added, and features like cupolas were fitted.

Specifications

Length:	4.75m
Width:	2.14m
Height:	1.95m
Weight:	7.9 tonnes
Crew:	3
Engine:	Maybach HL 57 TR 6-cylinder water-cooled 130hp petrol/gasoline
Armament:	2cm Kw.K.30 L/55 auto-cannon
Additional weapon:	7.92mm M.G.34 machine gun
Armour thickness:	5mm–16 mm
Max. road speed:	40km/h
Max. range on roads:	190km
Total built:	100

Panzer II Ausf.b (Sd.Kfz.121)

Chapter 6
Panzer II Ausf.c (Sd.Kfz.121)

The suspension on the Ausf.c was visually very different from that used on previous models. Five larger 55cm-diameter road wheels replaced the six small road wheels. The suspension was now a leaf-spring, crank-arm system. The long metal beam that ran along the road wheels was no longer needed and was removed. The new version of the front-drive wheel first introduced on the Ausf.b was kept. An additional track-return roller was added, bringing the total to four. The front track guard extension was now held together by clips.

This increased the total weight from 7.9 tonnes to 8.9 tonnes. This did not affect the tank's top speed as the engine was upgraded as well. It was fitted with a more powerful Maybach HL 62 TR 6-cylinder water-cooled 140hp petrol/gasoline engine.

The 2cm Kw.K.30 gun could fire three different shells. When fired against armour plate laid back at 30° from the vertical, the PzGr.39 armour-piercing shell could penetrate 23mm of armour at 100m and 14mm of armour at 500m. The PzGr.40 armour-piercing composite rigid shell could go through 40mm of armour at 100m and 20mm of armour at 500m. It could also fire 2cm Sprgr.39 high-explosive shells.

Early versions of the Panzer II changed shape over time as they were upgraded during their operational life. Additional armour was added, and features such as cupolas with periscopes were fitted. The bullet ricochet 'splash' plate and the dummy cone-shaped periscope in front of the commander's hatch were later removed. Turret ring guards were not present on this vehicle, but some had then backfitted. The additional armour added to the front hull glacis plates changed the look from a curved frontal armoured hull to an angular shape.

Specifications

Length:	4.81m
Width:	2.22m
Height:	1.99m
Weight:	8.9 tonnes
Crew:	3
Engine:	Maybach HL 62 TR 6-cylinder water-cooled 140hp petrol/gasoline
Armament:	2cm Kw.K.30 L/55 auto-cannon
Additional weapon:	7.92mm M.G.34 machine gun
Armour thickness:	5mm–16mm
Max road speed:	40km/h
Max. range on roads:	190km
Total built:	75

Panzer II Ausf.c (Sd.Kfz.121)

Chapter 7
Panzer II Ausf.A (Sd.Kfz.121)

The Panzer II Ausf.A was the final standardised version ready for mass production. The previous versions, Ausf.a/1, a/2, a/3, b and c, were all trial series developed to test new design elements. This is why a capital letter 'A' was used to denote the production version. Only minor internal changes were made: a new gearbox was fitted and the fuel pump, oil filter and cooler were relocated on the engine. The tank's electrical system was suppressed to try and stop it interfering with the AM radio reception and transmission.

The main visual difference between the Ausf.c and the Ausf.A was the introduction of a new driver's visor at the front of the tank. The large, flat, rectangular armoured vision port cover was now replaced with a V-shaped armoured visor that had a slit built into it. The two side visors used by the driver and radio operator were now of the same type. The early versions of the Panzer II Ausf.A did not have a turret ring guard bolted onto the superstructure at the front and rear of the turret ring.

The 2cm Kw.K.30 gun could fire three different shells. When fired against armour plate laid back at 30° from the vertical, the PzGr.39 armour-piercing shell could penetrate 23mm of armour at 100m and 14mm of armour at 500m. The PzGr.40 armour-piercing composite rigid shell could go through 40mm of armour at 100m and 20mm of armour at 500m. It could also fire 2cm Sprgr.39 (high-explosive) shells.

Panzer II tanks were designed to support infantry attacks. They would subdue enemy machine guns that could slow down and stop an attack. As they were impervious to rifle and machine-gun fire they could get close to their target and knock it out. Their tracks enabled them to cross most terrain and cross barbed-wire entanglements. Their speed would make them a difficult target for anti-tank guns to hit.

Specifications

Length:	4.81m
Width:	2.22m
Height:	1.99m
Weight:	8.9 tonnes
Crew:	3
Engine:	Maybach HL 62 TR 6-cylinder water-cooled 140hp petrol/gasoline
Armament:	2cm Kw.K.30 L/55 gun
Additional weapon:	7.92mm M.G.34 machine gun
Armour thickness:	5mm–16mm
Max. road speed:	40km/h
Max. range on roads:	190km
Total built:	210

Panzer II Ausf.B (Sd.Kfz.121)

Chapter 9
Panzer II Ausf.C (Sd.Kfz.121)

The Ausf.C was ordered to keep the factories busy until the Panzer III tank was ready for mass production. The only visible difference was a new type of improved vision port. It had two conical beaded bolts on the faceplate and two large bolts above and below it to keep the 50mm bulletproof glass in place. It was still armed with a 2cm Kw.K.30 L/55 main gun that could fire both armour-piercing and high-explosive shells. The turret was also fitted with a 7.92mm M.G.34 machine gun. The tanks were fitted with a two-piece commander's hatch. The driver's side vision ports were fitted with a bullet splash guard. The turret ring was protected by a triangular front and rear turret guard.

Early versions of the Panzer II changed shape over time as they were upgraded during their operational life. Additional armour was added to the front of the tank's hull and turret in 1940. On 23 October 1940, instructions were given that a commander's cupola was to be backfitted on the Panzer II Ausf.c, Ausf.A, Ausf.B and Ausf.C. The bullet ricochet 'splash' plate and the dummy cone-shaped periscope that used to be in front of the commander's hatch were removed. The additional armour added to the front hull glacis plates changed the look from a curved frontal armoured hull to an angular shape.

The Panzer II light tanks were first issued to panzer units in the spring of 1936. They were used for reconnaissance roles during the invasion of Belgium, Holland and France. During the invasion of Russia, Operation *Barbarossa*, which started on 22 June 1941, most panzer units had a platoon of Panzer II tanks for scouting reconnaissance missions. The platoons were withdrawn in 1942 and tanks were phased out from panzer regiments in 1943. They were still used for internal security work until the end of the war.

Specifications

Length:	4.81m
Width:	2.22m
Height:	1.99m
Weight:	8.9 tonnes
Crew:	3
Engine:	Maybach HL 62 TR 6-cylinder water-cooled 140hp petrol/gasoline
Armament:	2cm Kw.K.30 L/55 auto-cannon
Additional weapon:	7.92mm M.G.34 machine gun
Armour thickness:	5mm–16mm
Max. road speed:	40km/h
Max. range on roads:	190km
Total built:	364

Panzer II Ausf.C (Sd.Kfz.121)

Chapter 10
Panzer II Ausf.D (Sd.Kfz.121) and Ausf.E

The leaf-spring suspension on the Panzer II Ausf.c, Ausf.A, Ausf.B and Ausf.C tanks was found to have a limited life span of 1,500–2,500km before it needed changing. A new torsion-bar suspension system with larger road wheels and a different drive and idler-wheel were introduced on the Panzer II Ausf.D and Ausf.E. It was designed by Maschinenfabrik Augsburg-Nürnberg (M.A.N.). No track-return rollers were used. The seven Ausf.E hulls had different wheels. They were used for trials, never as combat tanks, as no turret or superstructure was fitted to them. They were converted to Panzer II (Flamm) flame-throwing tanks.

A new Maybach HL 62 TRM engine and a new Maybach 7-speed (1 reverse) S.R.G 14479 transmission enabled this heavier Panzer II Ausf.D tank to reach top speeds of 55km/h. Fuel tanks were moved into the engine compartment. The rear engine deck was completely changed. The armoured deck now covered the width of the tank and had two large split hatches in it.

One of the significant differences was that the radio operator now has his own armoured forward vision port and hatch at the front of the tank. The triangular aerial support on the left of the tank was removed, and the aerial was repositioned on the right side of the vehicle. There were no vertical bullet splash shields in front of the side, late-version, vision ports. There were conical-shaped bolts above and below the armoured side vision ports to hold in place the 50mm-thick bulletproof glass.

The front hull armour was now 30mm thick and of an angular rather than a curved design. The turret armour was still 14.5mm. It had a split hatch and dummy periscope cone and bullet splash guard in front of the hatch. Some Panzer II Ausf.D tanks that survived Poland and the invasion of France were converted into 7.62cm Pak 36(r) Marder II (Sd.Kfz.132) tank hunters following an order issued on 20 December 1941. Others were rebuilt as flame-throwing tanks.

Specifications

Length:	4.75m
Width:	2.14m
Height:	2.02m
Weight:	11 tonnes
Crew:	3
Engine:	Maybach HL 62 TRM 6-cylinder water-cooled 140hp petrol/gasoline
Armament:	2cm Kw.K.30 L/55 auto-cannon
Additional weapon:	7.92mm M.G.34 machine gun
Armour thickness:	5mm–30 mm
Max. road speed:	55km/h
Max. range on roads:	200km
Total built Ausf.D:	43
Total built Ausf.E:	7 (converted to flame-throwing tanks)

Panzer II Ausf.D (Sd.Kfz.121) and Ausf.E

Panzer II Ausf.D.

Panzer II Ausf.E converted to a flame-throwing tank.

Chapter 11
Panzer II Ausf.F (Sd.Kfz.121)

The Panzer II Ausf.F was built with the thicker 30mm armour on the front of the tank hull and 35mm armour on the front of the turret. (When measured, some front glacier plates were found to be 35mm thick.) The commander had a cupola with a periscope on the top of the turret rather than a split hatch. The side vision ports had vertical bullet splash guards in front of them and had two conical bolts above and below the visor to hold in place the 50mm bulletproof glass behind it.

The turret dummy periscope and commander's hatch bullet splash guard were not fitted. The turret ring was protected from bullet and shrapnel damage by a triangular-shaped guard welded to the top of the superstructure at the front and back. The turret was fitted with a rear stowage bin.

A fake armoured visor, made from aluminium alloy, was bolted onto the front of the hull to the right of the driver's vision port. This was done to distract enemy fire away from the driver. Most other parts used to build the tank were unchanged from previous models. It was still armed with a 2cm Kw.K.30 L/55 gun and 7.92mm M.G.34 machine gun.

The first seven Panzer II Ausf.F light tanks were completed in March 1941. Production stopped at the end of July 1942.

They were used mainly on the Eastern Front as reconnaissance tanks, but some Panzer II Ausf.F light tanks were sent to Libya as replacements. In the desert, they were issued to the 2nd Battalion, 5th Panzer Regiment, 21st Division (II.Abt/Pz.Rgt.5). These tanks had the size of the cooling air intake and exhaust holes increased and the radiator fan changed for a high-performance version so they could cope better with the hot desert temperatures. Late-production tanks built in 1942 had four posts fitted around the turret cupola to be used as a base for a Fla-MG anti-aircraft machine gun. The rear turret stowage bin does not seem to be fitted.

Specifications

Length:	4.81m
Width:	2.28m
Height:	2.15m
Weight:	9.5 tonnes
Crew:	3
Engine:	Maybach HL 62 TR 6-cylinder water-cooled 140hp petrol/gasoline
Armament:	2cm Kw.K.30 L/55 auto-cannon
Additional weapon:	7.92mm M.G.34 machine gun
Armour thickness:	5mm–35mm
Max. road speed:	40km/h
Max. range on roads:	190km
Total built:	509

Panzer II Ausf.F (Sd.Kfz.121)

Chapter 12
Panzer II Ausf.G (Sd.Kfz.121/1)

The Panzer II Ausf.G was fitted with a new, more powerful engine, the Maybach HL 45 water-cooled 150hp petrol/gasoline engine. It increased the tank's top road speed from the normal 40km/h of older versions to 65km/h. The 2cm automatic main gun was upgraded to the 2cm Kw.K.38 L/55 gun. A 7.92mm M.G.34 was also mounted in the turret. Both guns were stabilised, for the first time, to help increase accuracy when they were fired on the move.

The new overlapping torsion-bar suspension system's five large road wheels were the most identifiable feature of this tank. No track-return rollers were fitted. This enabled only a short length of track to be in contact with the ground, which resulted in exceptional manoeuvrability as it had a small turning circle. The first and last torsion bars on each side of the tank had shock absorbers attached to dampen down the impact of bumps at speed. During driving trials, several different transmission gearboxes were tested.

The front hull and front turret armour were now 30mm thick. The front of the tank was angled rather than curved. The side armour of the hull and turret was 14.5mm thick. The side vision ports were of the new style with two conical bolts above and below the port to hold the 50mm bulletproof glass in position behind it but no vertical bullet splash guard was fitted in front of the visor port. The front vision ports now had a protective armour plate that ran the width of the superstructure and was divided into three. The central one was fake to confuse enemy snipers. The armoured pipe in front of it was a steering brake fume exhaust.

A straight rather than curved turret ring bullet and shrapnel triangular splash guard was welded in front of the driver's and radio operator's hatches. One was not fitted behind the turret. Stowage boxes were not fitted to the rear of the turret at the factory. The engine deck covered the whole width of the tank. Photographs show them being used on the Eastern Front.

Specifications

Length:	4.24m
Width:	2.38m
Height:	2.05m
Weight:	10.5 tonnes
Engine:	Maybach HL 45 P water-cooled 150hp petrol/gasoline
Crew:	3
Main gun:	2cm Kw.K.38 L/55 auto-cannon
Other weapons:	7.92mm M.G.34(P) machine gun
Armour:	5.5mm–30mm
Max. road speed:	65km/h
Max. range on roads:	200km
Total built:	45

Panzer II Ausf.G (Sd.Kfz.121/1)

Chapter 13
Panzer II Ausf.J (Sd.Kfz.121)

Instructions were given to M.A.N. and Daimler-Benz to build a strengthened Panzer II tank. The frontal armour on the hull and turret was increased from 30mm to 80mm thick. The sides and rear of the turret and hull were increased from 14.5mm to 50mm thick. It was armed with a 2cm Kw.K.38 L/55 gun and a 7.92mm M.G.34 machine gun in the turret. The commander had a cupola on top of the turret.

The weight of the Panzer II was increased to 17.4 tonnes. It was powered by a Maybach HL 45 150hp petrol/gasoline engine and because of the weight only had a top road speed of 31km/h. This badly effected its ability to act as a scout car as speed was essential although its survivability prospects had increased.

It had the same overlapping torsion-bar suspension system as used on the Panzer II Ausf.G with five large road wheels and no track-return rollers. The most identifying feature of this design was the central gap in the track guard on both sides of the tank which enabled access to the crew's new round entrance and escape hatch. They had to be repositioned to this location as there was no longer any room above the heads of the driver and radio operator because of the increase in the thickness of the frontal armoured plate. Both sides also had a circular armoured cover around the side hull vision port. There was no turret ring protection shield at the rear of the turret but there was one welded to the top of the hull in front of the turret.

By the end of December 1942, a total of 30 Panzer II Ausf.J strengthened tanks had been built. They saw combat on the Eastern and Western fronts.

Specifications

Length:	4.24m
Width:	2.38m
Height:	2.05m
Weight:	17.4 tonnes
Engine:	Maybach HL 45 water-cooled 150hp petrol/gasoline
Crew:	3
Main gun:	2cm Kw.K.38 L/55 auto-cannon
Other weapons:	7.92mm M.G.34 machine gun
Armour:	5.5mm–30mm
Max. road speed:	31km/h
Max. range on roads:	175km
Total built:	30

Panzer II Ausf.J (Sd.Kfz.121)

Chapter 14
Panzerspähwagen II (2cm) 'Luchs' (Sd.Kfz.123)

In 1938, the German companies Maschinenfabrik Augsburg Nürnberg (M.A.N.) and Daimler-Benz were awarded the contract to design a new version of the Panzer II light tank for reconnaissance missions. They had already produced a three-man Panzer II; M.A.N. worked on the hull and Daimler-Benz constructed the superstructure and turret. They then moved on to develop a four-man version that would become the Panzerspähwagen II (2cm) (Sd.Kfz.123), also known as the Panzer II Ausf.L 'Luchs' ('Lynx'). When translated into English, Panzerspähwagen and Panzerspaehwagen mean armoured car.

The first prototype hull was completed in July 1941. In June 1942, it was tested against two Czech-built light tanks, the Škoda T 15 and 38(t) n.a. tanks. The Luchs was found to be the better design, with a larger turret and better ground clearance. During the trials, the engine, clutch and transmission functioned without problems over different terrains.

The Maybach 180hp HL 66 P water-cooled petrol/gasoline engine had enough power to enable the tank to have a top road speed of 60km/h.

The front armour on the turret and hull was 30mm thick. The side and rear armour were 20mm thick. The turret was armed with a centrally mounted 2cm Kw.K.38 L/55 main gun with a 1.3m-long anti-aircraft gun barrel and a coaxial 7.92mm M.G.34(P) machine gun which had an armoured sleeve to protect the gun barrel. The gunner sat on the right of the turret, which was a different layout from most German turrets. The Maybach HL 66 P water-cooled 180hp petrol/gasoline engine produced enough power to give the tank a top road speed of 60km/h. Production of the 2cm Luchs began in September 1942 and finished on 7 January 1944; only 100 were built. They were used on the Eastern Front and on the Western Front in Normandy.

Specifications

Length:	4.63m
Width:	2.13m
Height:	2.21m
Weight:	11.8 tonnes
Crew:	4
Engine:	Maybach HL 66 P water-cooled 180hp petrol/gasoline
Armament:	2cm Kw.K.38 L/55 auto-cannon
Additional weapon:	7.92mm M.G.34 machine gun
Armour thickness:	5.5mm–30mm
Max. road speed:	60km/h
Max. range on roads:	260km
Total built:	100

Panzerspähwagen II (2cm) 'Luchs' (Sd.Kfz.123)

Chapter 15
Panzer 35(t)

The most noticeable difference between the Panzer 35(t) and the Panzer 38(t) was the suspension. The 35(t) had nine pairs of small road wheels, whereas the 38(t) had four large road wheels. It saw service in the early years of World War Two.

Before the German occupation of Czechoslovakia, Škoda produced a light tank called the LT vz.35. When it started in service with the German Army, it was called the Panzer 35(t). The German word for Czechoslovakia is 'Tschechoslowakei', which is why the letter 't' appears in brackets. It was armed with a Škoda 3.7cm Kw.K.34(t) L/40 gun which was adequate for tanks designed in 1934. There was a coaxial 7.92mm M.G.37(t) machine gun in the turret and another in the hull. The hull was mostly riveted.

The Germans made some modifications. They changed the three-coloured light internal communication method to an intercom system. The hull machine gunner also operated the radio. In the original design, only one man was in the turret. The commander had to look out for threats and possible targets, load and fire the main gun and give directions to the crew. This was not ideal. He had too much to do. The Germans added a fourth member to the crew who was responsible for loading the main gun and firing the turret machine gun.

The Panzer 35(t) tanks were used in the 1939 German invasion of Poland. They fought in the Battle of France and in the Balkans. On 22 June 1941, they took part in the invasion of the Soviet Union. They could knock out Red Army T-26 and BT-7 tanks, but by 1942 it was clear that their use was limited to a reconnaissance role, as they could not knock out the new Soviet T-34 and KV-1 tanks. Many were withdrawn from the front line. Some were converted into self-propelled guns while others were used for internal security. The Panzer 35(t) was used by the Romanian, Slovakian and Bulgarian armies on the Eastern Front.

Specifications

Length:	4.90m
Width:	2.06m
Height:	2.37m
Weight:	10.5 tonnes
Engine:	Škoda T-11/0 4-cylinder water-cooled 8,620cc 120hp petrol/gasoline
Crew:	3
Main gun:	3.7cm Škoda ÚV vz.34 (3.7cm Kw.K.34(t) L/40)
Other weapons:	2 x 7.92mm ZB vz.37 machine guns (7.92mm M.G.37(t))
Armour:	8mm–25mm
Max. road speed:	34km/h
Max. range on roads:	120km
Total built:	643

Panzer 35(t)

Chapter 16
Panzer 38(t)

Before the German occupation of Czechoslovakia, Praga-Škoda had designed and produced a competent light tank called the LT vz.38. When it started in service with the German Army, it was called the Panzer 38(t). The design had been altered to make the turret large so that the tank had room for a third crewman, the gunner, an intercom system and a German radio set.

It was armed with a Škoda 3.7cm Kw.K.38 L/47.8 gun which was adequate for tanks designed in 1938. There was a coaxial 7.92mm Zb53 machine gun in the turret and another in the hull. The hull was mostly riveted. The most distinctive feature of this tank's design were the four large road wheels. Four different versions, Ausf.A to Ausf.D, had been produced by November 1940 with minor changes. Its maximum armour was 30mm thick. Crew protection was increased on the next versions, Ausf.E and Ausf.F, to a maximum of 50mm armour on the front of the tank. The Ausf.G had the same armour protection, but most of the hull was now welded. The Ausf.S was an export order for Sweden that was confiscated.

They equipped front-line panzer divisions who praised their reliability. They were used in the 1939 invasion of Poland and in April 1940 to occupy Norway. They fought in the Battle of France and in the Balkans. On 22 June 1941, they took part in the invasion of the Soviet Union. One tank could knock out Red Army T-26 and BT-7 tanks, but by 1942 it was clear that their use was now limited to a reconnaissance role, as they could not knock out the new Soviet T-34 and KV-1 tanks. Many were withdrawn from the front line. Some were converted into self-propelled guns while others were used for internal security.

Specifications

	Ausf.A and B	Ausf.C and D	Ausf.E and F	Ausf.G and S
Length:	4.61m	4.61m	4.61m	4.61m
Width:	2.12m	2.14m	2.14m	2.14m
Height:	2.25m	2.25m	2.25m	2.25m
Weight:	9.4 tonnes	9.5 tonnes	9.85 tonnes	9.85 tonnes
Engine:	Praga TNHPS/II	Praga TNHPS/II	Praga TNHPS/II	Praga TNHPS/II
Crew:	4	4	4	4
Main gun:	3.7cm Kw.K. M38(t) L/47.8	3.7cm Kw.K. M38(t) L/47.8	3.7cm Kw.K. M38(t) L/47.8	3.7cm Kw.K. M38(t) L/47.8
Other weapons:	2 x 7.92mm M.G.37(t)	2 x 7.92mm M.G.37(t)	2 x 7.92mm M.G.37(t)	2 x 7.92mm M.G.37(t)
Armour:	8mm–25mm	8mm–40mm	8mm–50mm	8mm–50mm
Max. road speed:	42km/h	42km/h	42km/h	42km/h
Max. range on roads:	250km	250km	250km	250km
Total built:	150 (A) 110 (B)	110 (C) 105 (D)	275 (E) 250 (F)	316 (G) 90 (S)

Panzer 38(t)

Chapter 17
Panzer III Ausf.A–D (Sd.Kfz.141)

The easiest way to tell the difference between a Panzerkampfwagen Mark III and a Mark IV is to count the road wheels. The Mk III has six pairs of road wheels each side and the longer Mk IV has eight. Unfortunately, this simple guide does not take into account the early experimental trial versions that were used to test different track and suspension systems, among other features. The Panzer III was the main battle tank of the German Army during the early years of World War Two. It would later be replaced by the Panzer IV, Panther and Tiger tanks.

On 27 January 1934, authorisation was given to develop a 10-ton tank with a 3.7cm Kw.K.36 L/46.5 gun in the turret, code-named 'Zugfuehrerwagen' ('platoon leader's tank'), abbreviated to Z.W. Two trial tank hull were ordered from Daimler-Benz, and one from M.A.N. Two trial turrets were commissioned from Krupp and one from Rheinmetall. These orders were increased.

Each side of the Panzer III Ausf.A (Z.W.1) had five large road wheels with coil-spring suspension and two track-return rollers. Each side of the Panzer III Ausf.B (Z.W.3) had eight smaller road wheels with leaf-spring suspension in two groups and three track-return rollers. Each side of the Panzer III Ausf.C and D (Z.W.4) had eight road wheels with leaf-spring suspension in three groups and three track-return rollers. The design of the leaf-spring suspension system on the Ausf.C was different from the leaf-spring suspension system fitted on the Ausf.D. They were all powered by a Maybach HL 108 TR 250hp engine but had different transmission gearboxes.

The armour on the front, sides and rear of the tank hull was 14.5mm thick. The turret sides and rear were also 14.5mm thick. The gun mantle and front were 16mm thick. The tanks saw combat during the invasions of Poland and Norway. As newer, more heavily armoured versions of the Panzer III arrived in panzer regiments as replacements, the surviving tanks were sent to tank crew training schools.

Specifications

	Ausf.A	Ausf.B	Ausf.C	Ausf.D
Length:	5.80m	5.66m	5.85m	5.92m
Width:	2.81m	2.81m	2.82m	2.82m
Height:	2.36m	2.38m	2.41m	2.41m
Weight:	15 tonnes	16 tonnes	16 tonnes	16 tonnes
Engine:	Maybach HL 108 TR 250hp petrol/gasoline			
Crew:	5	5	5	5
Main gun:	3.7cm Kw.K.36 L/46.5			
Other weapons:	3 x 7.92mm M.G.34 machine guns			
Armour:	5mm–16mm (14.5mm front, side and rear hull)			
Max. road speed:	35km/h	35km/h	35km/h	35km/h
Max. range on roads:	165km	165km	165km	165km
Total built:	10	15	15	30

Panzer III Ausf.A–D (Sd.Kfz.141)

Panzer III Ausf.A (Sd.Kfz.141).

Panzer III Ausf.B (Sd.Kfz.141).

Panzer III Ausf.C (Sd.Kfz.141).

Panzer III Ausf.D (Sd.Kfz.141).

Chapter 18
Panzer III Ausf.E (Sd.Kfz.141)

The Ausf.E was the first version of the mass-produced Panzer III tanks and was very similar to the Ausf.F and Ausf.G, which differed in minor specifications. The previous versions had been used to test different suspension systems and other features. The Panzer III Ausf.E was fitted with torsion-bar suspension with six road wheels on individual swing axles. Three track-return rollers were positioned above the road wheels. It had a cone in front of the cupola on the turret. This was a fake gunner's periscope to draw enemy fire. The rear of the tank was not fitted with a smoke grenade rack.

The Panzer III Ausf.E was fitted with the Maybach HL 120 TR 265hp engine. It was slightly more powerful than the one used in earlier versions of the Panzer III tank. The Ausf.F and Ausf.G were fitted with the Maybach HL 120 TRM 285hp engine, which had a different magneto and modified cooling system.

The armour on the Panzer III Ausf.E tank was 30mm thick on the driver's front plate, hull front, superstructure and hull side. The armour on the angled front glacis and upper glacis plates was 25mm thick. The hull rear was 20mm thick on the Ausf.E. The front, side and rear turret armour was 30mm thick. The gun mantlet was 30mm thick, as was the armour on the cupola.

The 3.7cm Kw.K.36 L/46.5 tank gun had a length of 1,716mm from the muzzle to the back of the breech. It had a rate of fire of up to 20 rounds per minute. This was achieved by having a semi-automatic breech which opened shortly before the end of the recoil, allowing for the ejection of the spent casing. The breech had to be opened by hand before the first shot, but it closed by itself when a round was loaded. Its PzGr.18 AP shells could penetrate 34mm-thick armour laid at a 30° angle at a range of 100m, 29mm at 500m and 22mm at 1km. This was adequate to deal with the threats it faced in 1939.

A few Panzer III Ausf.E tanks saw combat in Poland in 1939. They were used in the invasions of Holland, Belgium and France in May 1940. These tanks were upgraded during their combat life with different guns (5cm Kw.K.38 L/42) turrets and more armour. They were used on the Eastern Front and in North Africa.

Specifications

Length:	5.38m
Width:	2.91m
Height:	2.50m
Weight:	19.5 tonnes
Engine:	Maybach HL 120 TR 265hp petrol/gasoline
Crew:	5
Main gun:	3.7cm Kw.K.36 L/46.5 or 5cm Kw.K.38 L/42
Other weapons:	3 x 7.92mm M.G.34 machine guns (the 5cm gun turret only had one coaxial machine gun)
Armour:	10mm–30mm
Max. road speed:	40km/h
Max. range on roads:	165km
Total built:	96

Panzer III Ausf.E (Sd.Kfz.141)

Chapter 19
Panzer III Ausf.F (Sd.Kfz.141)

The Panzer III Ausf.F tank was very similar to the Ausf.E and Ausf.G. The Panzer III Ausf.E was fitted with torsion-bar suspension with six road wheels on individual swing axles. Three track-return rollers were positioned above the road wheels.

A turret ring deflector guard was added to the roof of the hull in front of the turret. The dummy periscope, designed to draw sniper fire, was removed from in front of the commander's cupola on later-built turrets. Some early ones still had it. A Nebelkerzenabwurfvorrichtung smoke grenade rack was added to the rear of the tank hull. Two armoured brake vents were fitted to the front upper glacis plate.

It was fitted with the 285hp HL 120 TRM petrol/gasoline engine which had a different magneto and modified cooling system from the HL 120 TR 250hp engine fitted on the Ausf.E.

The armour on the Panzer III Ausf.F tank was 30mm thick on the driver's front plate, hull front, superstructure and hull side. The armour on the angled front glacis and upper glacis plates was 25mm thick. The hull rear was 20mm thick on the Ausf.F. The turret armour front, side and rear was 30mm thick. The gun mantlet was 30mm thick and so was the armour on the cupola.

The 3.7cm Kw.K.36 L/46.5 tank gun had a length of 1,716mm from the muzzle to the back of the breech. It had a rate of fire of up to 20 rounds per minute. This was achieved by having a semi-automatic breech which opened shortly before the end of the recoil, allowing for the ejection of the spent casing and the quick loading of the next shell.

Late Panzer III Ausf.F tanks were fitted with 5cm Kw.K.38 L/42 guns. An armoured vent was fitted to the roof of the turret. Armoured vents were fitted to the rear engine deck to enable it to cope with the dust and heat of the North African desert. Stowage boxes were mounted on the rear of the turret on late-production tanks, and some were retrofitted later.

Specifications

Length:	5.38m
Width:	2.91m
Height:	2.50m
Weight:	19.5 tonnes
Engine:	Maybach HL 120 TRM 285hp petrol/gasoline
Crew:	5
Main gun:	3.7cm Kw.K.36 L/46.5 or 5cm Kw.K.38 L/42
Other weapons:	3 x 7.92mm M.G.34 machine guns (5cm gun turret only had one coaxial machine gun)
Armour:	10mm–30mm
Max. road speed:	40km/h
Max. range on roads:	165km
Total built:	435

Panzer III Ausf.F (Sd.Kfz.141)

Chapter 20
Panzer III Ausf.G (Sd.Kfz.141)

The Panzer III Ausf.G was produced between March 1940 and early 1941. It was very similar to the Ausf.E and Ausf.F with minor differences in specifications. The Panzer III Ausf.G was fitted with torsion-bar suspension with six road wheels on individual swing axles. Three track-return rollers were positioned above the road wheels. The front return roller was moved forward on later versions.

It had a turret ring deflector guard on the front of the hull superstructure. It had a smoke grenade launcher mounted to the rear of the tank hull. Two armoured brake vents were fitted to the front upper glacis plate. Armoured vents were fitted to the turret roof and to the rear of the engine deck. The driver's visor was changed from a partitioned visor to a pivoting visor. A 90mm-thick glass block protected the eyes of the driver. He was also provided with a new K.F.F.2 double periscope that used prisms instead of mirrors and gave a 50° field of view. The commander had a new, improved cupola with five independently adjustable vision slots.

The armour on the Panzer III Ausf.G tank was 30mm thick on the driver's front plate, hull front, superstructure and hull side. The armour on the angled front glacis and upper glacis plates was 25mm thick. The hull rear was now 30mm thick on the Ausf.G, an upgrade from the old standard of 20mm. The front, side and rear turret armour was 30mm thick. The gun mantlet was 30mm thick and so was the armour on the cupola. Additional 30mm-thick armour plates were later bolted onto the front and rear of the tank.

The first Ausf.G tanks were armed with a 3.7cm Kw.K.36 L/46.5 tank gun. Some took part in the invasion of Holland, Belgium and France in May 1940. Starting from July 1940, new Panzer III Ausf.G tanks were armed with the 5cm Kw.K.38 L/42 gun. Different factories started fitting the new gun at different times. They were used on the Eastern Front and in North Africa. These tanks were upgraded during their combat life with different guns, turrets and more armour. Rear turret stowage boxes were sometimes fitted later.

The factory-painted dark grey (dunkelgrau RAL 46) and dark brown (dunkelbraun RAL 45) camouflage pattern was discontinued by an order dated 31 July 1940. The tanks were just painted dunkelgrau after that date. Those going to North African were painted dark yellow (dunkelgelb).

Specifications

Length:	5.38m
Width:	2.91m
Height:	2.50m
Weight:	19.5 tonnes
Engine:	Maybach HL 120 TRM V-12 285hp petrol/gasoline
Crew:	5
Main gun:	3.7cm Kw.K.36 L/46.5 or 5cm Kw.K.38 L/42
Other weapons:	3 x 7.92mm M.G.34 machine guns (the 5cm gun turret only had one coaxial machine gun)
Armour:	10mm–30mm
Max. road speed:	40km/h
Max. range on roads:	165km
Total built:	594

Panzer III Ausf.G (Sd.Kfz.141)

Chapter 21
Panzer III Ausf.H (Sd.Kfz.141)

The Panzer III Ausf.H was the first version of the tank to be specifically designed with a turret fitted with the 5cm Kw.K.38 L/42 tank gun and with 60mm of frontal armour, rather than having these specifications added later in an upgrade program. The tanks started to be delivered in late 1940 and early 1941. They had the new-style cupola. The front return roller was moved forward.

The 5cm Kw.K. L/42 tank gun was semi-automatic; the breech block remained open after firing to enable the next round to be loaded quicker. Its standard armour-piercing shell could penetrate 55mm of armour laid at an angle of 30 degrees at a range of 100m, 46mm at 500m and 36mm at a range of 1km. The turret only had one coaxial 7.92mm M.G.34 machine gun, and another M.G.34 was mounted in the hull.

The tank was still powered by the Maybach HL 120 TRM V-12 water-cooled 285hp petrol/gasoline engine which gave it a top road speed of 42km/h. Two armoured brake vents were fitted to the front of the hull armour.

The 60mm-thick armour on the hull front and rear was constructed by bolting or welding two 30mm armour plates together. The side armour was 30mm thick, and the 87° angled front glacis plate was 25mm thick. The angled armour on the front, rear and sides of the turret was 30mm thick. The curved gun mantle was 35mm thick. The turret had an armoured ventilation fan. Tanks going to North Africa were fitted with armoured vents on the engine deck. Rear turret stowage bins were mounted later.

Because of the increase in weight, wider wheels and tracks were introduced. New front drive wheels and rear idler-wheels were fitted as well as a different shock absorber. As a result of supply problems, some of the early Ausf.H tanks were fitted with shock absorbers and wheels used on the Ausf.G.

Specifications

Length:	5.38m
Width:	2.95m
Height:	2.50m
Weight:	21.5 tonnes
Engine:	Maybach HL 120 TRM V-12 water-cooled 285hp petrol/gasoline
Crew:	5
Main gun:	5cm Kw.K.38 L/42
Other weapons:	2 x 7.92mm M.G.34 machine guns
Armour:	10mm–60mm
Max. road speed:	40km/h
Max. range on roads:	165km
Total built:	286

Panzer III Ausf.H (Sd.Kfz.141)

49

Chapter 22
Panzer III Ausf.J (Sd.Kfz.141) and Ausf.L (Sd.Kfz.141/1)

The Panzer III Ausf.J was very similar to the Panzer.III Ausf.G. It was built with a turret fitted with a 5cm Kw.K.38 L/42 tank gun. It had similar armour thickness and was powered by the same Maybach HL 120 TRM V-12 water-cooled 285hp petrol/gasoline engine.

The hull was lengthened to create better engine compartment ventilation and tow eyes. There was a more significant gap between the rear road wheel and the idler-wheel. The design of the armoured front brake vents was changed. The turret was fitted with an armoured extractor fan on the roof.

The armour thickness at the hull front, upper hull front and rear of the tank was now 50mm. The front glacis was 25mm thick and 30mm armour was used on the hull sides, lower hull rear and front. The armour on the front, sides and rear of the turret was 30mm thick. The rounded gun mantle was 50mm thick. In the spring of 1941, additional armour plate was added internally to the front of the turret, increasing it to a maximum of 57mm in places.

The 5cm KampfwagonKanone (Kw.K. – tank gun) had a length of 2,100mm (L/42) from the muzzle to the back of the breech. It had a rate of fire of up to 20 rounds per minute. This was achieved by having a semi-automatic breech which opened before the end of the recoil and ejected the spent casing, allowing for the quick loading of the next shell.

From December 1941, the 5cm Kw.K. L/60 tank gun had a length of 3,000mm. It started to be fitted instead of the 5cm Kw.K. L/42 gun as stocks arrived in factories. The tanks were renamed Panzer III Ausf.L and those sent to North Africa had armoured vents fitted on the rear engine deck. In April 1941, stowage bins started to be fitted to the rear of the turret. It is not possible to identify the exact number of tanks manufactured with the L/42 or L/60 guns as no distinction was made in the records.

Specifications

	Ausf.J	Ausf.L
Length:	5.49m	6.16m
Width:	2.95m	2.95m
Height:	2.50m	2.50m
Weight:	21.6 tonnes	22.5 tonnes
Engine:	Maybach HL 120 TRM V-12 water-cooled 285hp petrol/gasoline	
Crew:	5	5
Main gun:	5cm Kw.K.38 L/42	5cm Kw.K.38 L/60
Other weapons:	2 x 7.92mm M.G.34 machine guns	
Armour:	16mm–57 mm	16mm–57mm
Max. road speed:	40km/h	40km/h
Max. range on roads:	155km	155km
Total built:	approx. 1,521 L/42 and approx. 1,021 L/60	

Panzer III Ausf.J (Sd.Kfz.141) and Ausf.L (Sd.Kfz.141/1)

Chapter 23
Panzer III Ausf.M (Sd.Kfz.141/1) and Ausf.N (Sd.Kfz.141/2)

Contracts were placed for the Panzer III Ausf.M in February 1942. It had the same features as the Ausf.L, but was fitted with deep-wading equipment. Both versions of the tank were powered by a Maybach HL 120 TRM V-12 water-cooled 285hp petrol/gasoline engine. It was armed with the same 5cm Kampfwagenkanone 39 L/60 (5cm Kw.K.39 L/60) tank gun, with a length of 3,000mm, as used on the Ausf.L. The longer barrel gave the gun a higher velocity and penetration power than the shorter 5cm Kw.K. L/42, but it had problems penetrating the frontal armour of the T-34 and KV-1 at long range.

The Panzer III Ausf.N mounted a short-barrel 7.5cm Kampfwagenkanone 37 L/24 (7.5cm Kw.K.37 L/24) tank gun that was previously used on the Panzer IV. It was a low-velocity tank gun that was designed to fire mainly high-explosive shells. If it had to engage armoured vehicles in combat, it could fire the Panzergranate armour-piercing shell, but it was only effective at short ranges. Later on in the war, crews had the option to load the new 7.5cm HL-granaten 39 hollow-charge high-explosive anti-tank HEAT projectiles which had a greater effect against tank armour. The Panzer III Ausf.N was increasingly used in the infantry support role once the 75mm long-barrelled Panzer IV, Panther and 88mm-armed Tiger tank entered service.

Starting in May 1943, Schürzen 5mm skirt armour plates were mounted on the hull side and there were 10mm plates on the turret to prevent the Soviet 14.5mm anti-tank rifle penetrating the side armour of the Panzer III. Draftgeflecht metal mesh screens were also trialled. They were both as effective as each other, but the Schürzen skirt armour plates entered production as it would have taken too long to develop the support hangers for the metal mesh screens.

Specifications

	Ausf.M	Ausf.N
Length:	6.41m	5.49m
Width:	2.95m	2.95m
Height:	2.50m	2.50m
Weight:	22.5 tonnes	23 tonnes
Engine:	Maybach HL 120 TRM V-12 water-cooled 285hp petrol/gasoline	
Crew:	5	5
Main gun:	5cm Kw.K.39 L/60	7.5cm Kw.K.37 L/24
Other weapons:	2 x 7.92mm M.G.34 machine guns	
Armour:	16 mm–60 mm	16 mm–60 mm
Max. road speed:	40km/h	40km/h
Max. range on roads:	155km	155km
Total built:	250 approx.	614–750 approx.

Panzer III Ausf.M (Sd.Kfz.141/1) and Ausf.N (Sd.Kfz.141/2)

Panzer III Ausf.M.

Panzer III Ausf.M.

Panzer III Ausf.N.

Panzer III Ausf.N.

Chapter 24
Panzer IV Ausf.A (Vs.Kfz.622)

The long version of the tank's name is Panzerkampfwagen IV (7,5cm) (Vs.Kfz.622) Ausführung A (1. Serie/B.W.). Production of the Panzer IV Ausf.A started in November 1937 and ended in June 1938, after 35 vehicles had been completed. The Ausf.A was very similar to the B.W.I prototype and had an eight-road-wheel suspension which borrowed only a few unchanged parts from its predecessor. The whole vehicle consisted of four sub-assemblies: the turret, the superstructure front, the superstructure centre and the superstructure rear and lower hull. They were all bolted together.

The early Maybach HL 108 TR 230hp petrol/gasoline engine was located in the rear and separated from the crew compartment by a bulkhead. The V-12 engine enabled the vehicle to achieve a top road speed of 32.4km/h and a range of approximately 210km. It was connected to the ZF S.S.G.75 transmission with five forward gears and one reverse gear.

Each side of the hull featured eight rubber-rimmed road wheels mounted in pairs on leaf-springed bogeys bolted to the lower hull sides, a drive sprocket at the front and an idler-wheel at the rear with four rubber-rimmed track-return rollers. Track tension was provided by the adjustable idler-wheel.

The gunner was located to the front left of the commander, aiming through a telescopic sight, the T.Z.F.5b (Turm-Ziel-Fernrohr 5b – turret gunnery sight 5b), with a magnification of 2.5 and 25 degrees field of view (444m at a distance of 1,000m). The gunner fired the main gun electrical by means of a pistol grip attached to the handwheel of the turret traverse and the coaxial machine gun with a foot lever.

One unique feature of the Ausf.A was a foldable anti-aircraft mount for a machine gun attached to the left side of the superstructure, providing the crew with limited AA cover during rest. The Panzer IV Ausf.A had a total combat-loaded weight of 18 tonnes and its armour thickness ranged from 8mm to 16mm (14.5mm on the hull front).

Specifications

Length:	5.92m
Width:	2.83m
Height:	2.68m
Weight:	18 tonnes
Engine:	Maybach HL 108 TR V-12 230hp petrol/gasoline
Crew:	5
Main gun:	7.5cm Kw.K.37 L/24
Other weapons:	2 x 7.92mm M.G.34 machine guns
Armour:	8mm–16mm (14.5mm hull front)
Max. road speed:	32.4km/h
Max. range on roads:	210km
Total built:	35

Panzer IV Ausf.A (Vs.Kfz.622)

Chapter 25
Panzer IV Ausf.B (Vs.Kfz.622)

The long version of the tank's name is Panzerkampfwagen IV (7,5cm) (Vs.Kfz. 622) Ausführung B (2.Serie/B.W.). Krupp-Gruson completed 42 Panzer IV Ausf.B tanks between May and October 1938, a further three of the contract for a total of 45 vehicles were not completed because of problems with critical parts. The main changes on the Ausf.B, when compared to the previous Ausf.A, were the thickening of the frontal armour to 30mm and a new Maybach HL 120 TR V-12 water-cooled 265hp petrol/gasoline engine connected to a six-speed SSG-76 transmission, offering a top road speed of 42km/h.

The driver's armoured front was fabricated from one piece but without a ball mount for the 7.92mm M.G.34 machine gun. Instead, a rectangular visor with an armoured flap was mounted in front of the radio operator. A circular pistol port protected by an armoured cover was fitted to the lower right of the visor. The radio operator could fire a 7.92mm M.G.34 machine gun, submachine gun or pistol through this opening to fend off enemy infantry.

A new Fahrersehklappe-30 (driver's visor No.30) replaced the older, smaller version that was fitted to the Ausf.A. It consisted of two movable sliders mounted above and below a rectangular opening protected by 12mm-thick bulletproof glass. Both sliders could be closed to protect the opening from heavy enemy fire. In this case, the driver could observe the area in front of his tank through a telescope with two small openings located just over the driver's visor.

The split-hatches for both driver and radio operator were replaced with single-piece hatches opening to the front of the vehicle. A slightly smaller, better-armoured split-hatch commander's cupola with only five vision slits, protected by bulletproof glass, replaced the drum-shaped cupola of the earlier Ausf.A version. The vision slits of the cupola could be protected by two armoured sliders mounted above and below the opening. Due to the increased armour strength, the weight of the Ausf.B increased to 18.5 tonnes.

Specifications

Length:	5.92m
Width:	2.83m
Height:	2.68m
Weight:	18.5 tonnes
Engine:	Maybach HL 120 TR V-12 265hp petrol/gasoline
Crew:	5
Main gun:	7.5cm Kw.K.37 L/24
Other weapons:	2 x 7.92mm M.G.34 machine guns
Armour:	8mm–30mm (30mm hull front)
Max. road speed:	42km/h
Max. range on roads:	210km
Total built:	42

Panzer IV Ausf.D (Sd.Kfz.161)

Chapter 28
Panzer IV Ausf.E (Sd.Kfz.161)

The full name of this version of the Panzer IV was Panzerkampfwagen IV (7,5cm) Ausführung E (6.Serie/B.W.) (Sd.Kfz. 161). Of the 206 Panzer IV Ausf.E medium tanks ordered, a total of 200 were completed between October 1940 and April 1941. Of the six remaining vehicles, four hulls were used to construct armoured-vehicle-launched bridge tanks (AVLB) and the two others were modified with a Schachtellaufwerk (box running gear) and participated in extensive trials.

A new drive sprocket without side holes and improved road wheels with new hubcaps for improved lubrication were mounted on the Ausf.E. The two hatches in the vehicle front offering entrance to the steering brakes were embedded in the armour plating. While the driver's front remained the same as on the previous Panzer IV Ausf.D, the Fahrersehklappe-30 driver's visor was changed to the version already used on the Panzer III Ausf.G. An armoured smoke grenade launcher was mounted on the left side of the rear engine deck. A new, better-armoured split-hatch commander's cupola with five vision slits, the same as already used on the Panzer III Ausf.G, was mounted on the turret roof.

The turret rear was changed to a single plate without the overhang of the previous versions. It had a single circular signal gun-barrel opening on the left side of the turret roof. An exhaust fan with an armoured cover that had been located on the right of the turret roof was now moved further towards the main gun.

The frontal armour of the Ausf.E was increased to 50mm and many, but not all, Ausf.E tanks were up-armoured with additional 30mm appliqué armour bolted or welded to the driver's front and vehicle bow. Some had 20mm appliqué armour bolted or welded to the sides. The improvements added to the Ausf.E increased the vehicle's weight to 22 tonnes.

Specifications

Length:	5.92m
Width:	2.84m
Height:	2.68m
Weight:	22 tonnes
Engine:	Maybach HL 120 TRM V-12 265hp petrol/gasoline
Crew:	5
Main gun:	7.5cm Kw.K.37 L/24
Other weapons:	2 x 7.92mm M.G.34 machine guns
Armour:	8mm–60mm (60mm hull front)
Max. road speed:	42km/h
Max. range on roads:	210km
Total built:	200

Panzer IV Ausf.E (Sd.Kfz.161)

Chapter 29
Panzer IV Ausf.F (Sd.Kfz.161) and F2 (Sd.Kfz.161/1)

The full name of this early version of the Panzer IV was Panzerkampfwagen IV (7,5cm) Ausführung F (7.Serie/B.W.) (Sd.Kfz. 161). The Ausf.F was a landmark in the Panzer IV evolution and development. The early model, 'F', called 'F1' when the next model appeared, was the last of the 'short' versions. The front bow plate appliqué was now replaced by a full 50mm-thick armoured plate. Side armour and turret thickness were raised to 30mm. The total weight rose to more than 22 tonnes, which triggered other modifications, like larger track links (from 380mm to 400mm) to reduce ground pressure, and both the idler-wheel and front drive sprockets were modified in turn. Before its replacement in March, 464 units of the F1 were produced. The last 42 were altered to the new F2 standard.

Even equipped with the AP Panzergranate, the low-velocity gun of the Panzer IV was inadequate against well-armoured tanks. In the context of the upcoming campaign in Russia, some decision had to be made, which also concerned the long-awaited major upgrade of the Panzer III. The now largely available Pak 38 L/60, which already had been proved lethal, was supposed to be mounted in the turret of the Panzer IV by Krupp. In November 1941, the prototype was ready, and production was scheduled to start on the F2 standard. However, with the first encounters of Russian KV-1s and T-34s, the 50mm gun, also produced for the Panzer III, was dropped in favour of a new, more powerful model built by Rheinmetall, which was based on the 7.5cm Pak 40 L/46.

This led to the development of the 7.5cm Kw.K.40 L/43 tank gun. It was fitted with a muzzle brake to help reduce its recoil and extend the operational life of the gun barrel. Muzzle velocity, with the Panzergranade 39, topped at 990m/sec. It could penetrate 77mm of armour at up to 1,850m. After the first prototype was produced by Krupp, in February 1942, production of the F2 started. By July 1942, 175 had been delivered. However, in June 1942, the F2 was renamed Ausf.G, and further modifications were applied on the production line, but both types were known to the Waffenamt as the Sd.Kfz.161/1. Some nomenclatures and reports also speak of it as the F2/G version.

Specifications

	Ausf.F	Ausf.F2
Length:	5.92m	6.63m
Width:	2.88m	2.88m
Height:	2.68m	2.68m
Weight:	22.3 tonnes	23.6 tonnes
Engine:	Maybach HL 120 TRM V-12 265hp petrol/gasoline	
Crew:	5	5
Main gun:	7.5cm Kw.K.37 L/24	7.5cm Kw.K.40 L/43
Other weapons:	2 x 7.92mm M.G.34 machine guns	
Armour:	8mm–50 mm	8mm–80mm
Max. road speed:	42km/h	42km/h
Max. range on roads:	210km	210km
Total built:	464	42

Panzer IV Ausf.F (Sd.Kfz.161) and F2 (Sd.Kfz.161/1)

Panzer IV Ausf.F.

Panzer IV Ausf.F.

Panzer IV Ausf.F2.

Panzer IV Ausf.F2.

Chapter 30
Panzer IV Ausf.G (Sd.Kfz.161/1)

The long name of the final version of the Panzer IV was Panzerkampfwagen IV (7,5cm) Ausführung G (8.Serie/B.W.) (Sd.Kfz. 161/1). The Panzer IV Ausf.G was an improved F2 with armour modifications, including a weight-saving solution consisting of a progressive glacis side armour, which was thicker at the base. The frontal glacis received a new 30mm appliqué plate, giving a total of 80mm. This was largely sufficient against the Russian medium-velocity 76mm gun and the fearful 76.2mm anti-tank gun. At first, it was decided to bring only half production to this standard, but Adolf Hitler personally ordered, in January 1943, that the full output be upgraded, a decision that was well received by the crews.

However, the weight rose to 23.6 tonnes, further stressing the limited capacity of the hull and transmission. Both unit reports and mass-production requirements commanded further modifications. The turret vision port slits were eliminated, the engine ventilation and ignition at low temperatures were improved, and additional racks were fitted for spare road wheels and brackets for track links on the glacis. These acted as makeshift protection as well. A new headlight was installed, and the commander's cupola was up-armoured and modified.

The late-production versions, in March–April 1943, saw the introduction of side skirt armour (Schürzen) to the sides and turret, the latter equipped with smoke grenade launchers. Most importantly, they received the new 7.5cm Kw.K.40 L/48 gun, which had greater armour penetration power than the shorter-barrelled 7.5cm Kw.K.37 L/24 that was fitted on the early Ausf.F tanks. After 1,687 had been delivered by Krupp-Gruson, Vomag and Nibelungenwerke, including 412 of the up-gunned type, the production shifted towards the Ausf.H.

Specifications

Length:	6.63m
Width:	2.88m
Height:	2.68m
Weight:	23.6 tonnes
Engine:	Maybach HL 120 TRM V-12 265hp petrol/gasoline
Crew:	5
Main gun:	7.5cm Kw.K.40 L/43 (late production models had the 7.5cm Kw.K.40 L/48)
Other weapons:	2 x 7.92mm M.G.34 machine guns
Armour:	8mm–80mm (80mm hull front)
Max. road speed:	42km/h
Max. range on roads:	210km
Total built:	1,927

Panzer IV Ausf.G (Sd.Kfz.161/1)

Chapter 31
Panzer IV Ausf.H (Sd.Kfz.161/2)

The long name of the final version of the Panzer IV was Panzerkampfwagen IV (7,5cm) Ausführung H (9.Serie/B.W.) (Sd.Kfz. 161/2). The Ausf.H was equipped with the new long-barrelled high-velocity 7.5cm Kw.K.40 L/48 and was subsequently registered as the Sd.Kfz.161/2 by the ordnance department. Other modifications included simplifications to ease production, such as the removal of the hull side vision ports and, later, part sharing with the Panzer III. This was by far the biggest production of the type, with a total of approximately 2,322 machines, until its replacement by the Ausf.J in June 1944.

In December 1942, Krupp had received a request for a new version featuring all-sloped armour, which would have also required a new hull, transmission and probably engine as well, because of the added weight. However, production started with an upgraded version of the Ausf.G instead. A new headlight, a set of radios (FU2 and 5) and a new intercom were added. The Zahnradfabrik ZF SSG-76 transmission was fitted to cope with the increase in weight due to the front armour thickness being upgraded to 80mm, with no appliqué armour plates. Panzerschürzen skirt armour was added to the side of the tank, to protect the vulnerable area above the road wheels from Soviet 14.5mm anti-tank rifles, and to the turret sides and rear.

The Ausf.H now stood at 25 tonnes in battle order, and the maximum speed fell to 38km/h but only 25km/h in real combat conditions and far less on rough terrain. By the end of 1943, Zimmerit paste was factory-applied, new air filters were fitted, along with a turret anti-aircraft mount for an extra M.G.34 (Fliegerbeschussgerat), and modifications were made to the commander's cupola. The side and turret spaced armour were also factory-mounted.

Specifications

Length:	7.02m
Width:	2.88m
Height:	2.68m
Weight:	25 tonnes
Engine:	Maybach HL 120 TRM V-12 265hp petrol/gasoline
Crew:	5
Main gun:	7.5cm Kw.K.40 L/48
Other weapons:	2 x 7.92mm M.G.34 machine guns
Armour:	8mm–80mm (80mm hull front)
Max. road speed:	38km/h
Max. range on roads:	210km
Total built:	2,322 approx.

Panzer IV Ausf.H (Sd.Kfz.161/2)

Chapter 32
Panzer IV Ausf.J (Sd.Kfz.161/2)

The long name of the final version of the Panzer IV was Panzerkampfwagen IV (7,5cm) Ausführung J (10.Serie/B.W.) (Sd.Kfz. 161/2). The last type, the Ausf.J, began to roll off the factory line at Nibelungenwerke (at St Valentin, Austria) and Vomag, as Krupp was now involved with other tasks. It incorporated more mass-production oriented simplifications, which were rarely welcomed by the crews. A first example was the removal of the electric turret drive, traversing being done manually, which was sacrificed for an additional 200 litres of fuel capacity, raising the operational range to 300km. Tanks deployed to the Eastern front had to cover long distances. The further they could drive before having to refuel helped ease problems in the supply chain. Panzerschürzen skirt armour was added to the side of the tank, to protect the vulnerable area above the road wheels from Soviet 14.5mm anti-tank rifles, and to the turret sides and rear.

Other modifications included the removal of the turret visor, pistol ports and turret AA mount in favour of a Naehverteidigungswaffe mount. Zimmerit was not applied anymore, nor was the Schurzen, which was replaced by cheaper Thoma-type wire-mesh panels. The engine's radiator housing was also simplified. The drive train lost one return roller, and two Flammentoeter (flame-suppressing) mufflers were installed, as well as Pilze 2-ton crane mount sockets. More critically, the late Panzer III SSG 77 transmission was mounted, even though it was clearly overloaded.

Despite these sacrifices, the Ausf.J monthly deliveries were increasingly threatened by Allied bombings and the shortages it caused. About 2,970 to 3,420 (depending on sources) were built out of the planned of 5,000, including modified models sporting the Panther turret. All prototypes developed by 1942 were dropped in favour of the Panther. The hull was also used for some variants.

Specifications

Length:	7.02m
Width:	2.88m
Height:	2.68m
Weight:	25 tonnes
Engine:	Maybach HL 120 TRM V-12 265hp petrol/gasoline
Crew:	5
Main gun:	7.5cm Kw.K.40 L/48
Other weapons:	2 x 7.92mm M.G.34 machine guns
Armour:	8mm–80mm (80mm hull front)
Max. road speed:	38km/h
Max. range on roads:	210km
Total built:	between 2,970 and 3,420

Panzer IV Ausf.J (Sd.Kfz.161/2)

Chapter 33
Panzer IV mit Hydrostatischem Antrieb

In April 1945, US troops captured the Zahnradfabrik factory in Augsburg, north-west of Munich in southern Germany. They discovered the turretless hull of the prototype Panzer IV Ausf.G mit Hydrostatischem Antrieb tank. The turret was nearby.

The tank had an enlarged, curved rear engine compartment and smaller rear drive wheels. The tank was powered by the same Maybach HL 120 TRM 265hp petrol/gasoline engine fitted to most Panzer IV tanks. The reason for the redesigned engine compartment was to house a new transmission system.

It was called the Hydrostatischem Antrieb or Hydrostatic Drive, also known as the Thoma Drive. It was a hydraulic system that used oil as the liquid to change gears. Two oil pumps were installed behind the engine. They were connected directly to it. These powered two hydraulic motors, one on each side of the vehicle. A swash plate drive sent the power through a reduction gear into the newly added rear smaller drive wheels. They moved the tank tracks. They replaced the standard rear idler-wheels. The original toothed drive sprocket wheels at the front of the tank were replaced by a 78cm adjustable toothed idler-wheel. This tank was powered from the rear of the vehicle, not the front. The driver's steering leavers were replaced by a crescent-shaped steering wheel.

The Americans shipped the tank to the USA, where it underwent numerous tests. When these had been completed, it was put on show in an open field at the U.S. Army Ordnance Proving Grounds, Aberdeen, Maryland. In 2015, it was moved to the U.S. Army Center for Military History Storage Facility, Anniston, AL, USA. The staff gave it a long-winded official US Army designation: Tank, Medium, Full Track, Experimental Transmission, German Army, Steel, Tank, PzKpfw IV, 75mm Gun, German, 1945, World War II.

Specifications

Length:	5.41m
Width:	2.88m
Height:	2.68m
Weight:	25 tonnes
Engine:	Maybach HL 120 TRM 265hp petrol/gasoline
Crew:	5
Main gun:	7.5cm Kw.K.40 L/43
Other weapons:	7.92mm M.G.34/M.G.42
Armour:	15mm–65mm
Max. road speed:	42km/h
Max. range on roads:	210km
Total built:	1

Panzer IV mit Hydrostatischem Antrieb

Chapter 34
Panther Ausf.D (Sd.Kfz.171)

The first production Panther tank was the Ausf.D, not the Ausf.A. This confuses many people. It was followed by the Ausf.G. It was armed with a long-barrelled high-velocity 7.5cm Kw.K.42 L/70 gun that could knock out most Allied and Soviet tanks at long distances. It had an accurate, effective direct fire range of 1.1km–1.3km. With a good gun crew, it could fire six rounds a minute.

The upper front glacis plate armour was 80mm thick and angled at 55º. This meant that an enemy shell firing directly at the Panther from a head-on position would have to penetrate 139mm of armour plate due to the angle of the armour. The Tiger I tank only had 100mm of armour. This is a little-understood fact. The side armour was only 40mm thick. Starting in April 1943, Panzerschürzen skirt armour was added to the side of the Panther to protect the vulnerable area above the road wheels from Soviet 14.5mm anti-tank rifles.

The Panther's wide tracks and large interleaved road wheels resulted in lower ground pressure. This helped it traverse waterlogged, or deep-snow-covered rough terrain, providing better traction and mobility.

Design features were changed or deleted throughout the production run of each version of the Panther. Early Ausf.D turrets had a drum-shaped cupola, pistol ports in the side and rear armour, a loading hatch and no rain guard over the binocular gun-sight holes.

As with all new tank designs, there were mechanical reliability issues. Gradually, these problems were corrected. When the first batch of Panthers left the factory, they were painted Dunkelgrau (dark grey). In February 1943, all factories were instructed to paint all German armoured fighting vehicles Dunkelgelb, a dark sandy yellow. Each individual panzer unit then applied its own individual camouflage pattern.

Specifications

Length:	8.86m
Width:	3.27m
Height:	2.99m
Weight:	44.8 tonnes
Engine:	Maybach HL 210 P30 water-cooled 650hp petrol/gasoline (first 250 tanks) or Maybach HL 230 P30 V-12 water-cooled 700hp petrol/gasoline
Crew:	5
Main gun:	7.5cm Kw.K.42 L/70
Other weapons:	2 x 7.92mm M.G.34 machine guns
Armour:	16mm–80mm
Max. road speed:	55km/h
Max. range on roads:	200km
Total built:	842 approx.

Panther Ausf.D (Sd.Kfz.171)

Chapter 35
Panther Ausf.A (Sd.Kfz.171)

The hull used for the early production Panzer V Ausf.A was precisely the same as that used for the Ausf.D. This new batch of Panther tanks was given a new version designation, Ausf.A, because they were fitted with an improved turret.

Many features of the Ausf.D, like the drum-shaped commander's cupola and the thin rectangular 'letterbox' hull machine gun port, were still present on early-production Ausf.A Panthers produced between July and December 1943. They only changed mid-production and not at the same time. Other modifications were introduced during the production run. Ausf.D and Ausf.A tanks were also upgraded with different features once they had been issued to a panzer division.

In late November 1943, a ball mount (Kugelblende) with a spherical armoured guard was introduced. The radio operator could now see forward through the machine-gun sight. The forward-facing periscope was no longer fitted and the side periscope was repositioned 25mm further to the right.

The early-production Ausf.A turrets had three pistol ports: one on each side and one on the rear. To make production simpler and the armour stronger, the pistol ports were dropped from late-production Ausf.A turrets. Instead, a Nahverteidgungswaffe close-defence weapon was fitted to the roof of the tank to the right of the commander's cupola. It could fire a high-explosive grenade in the direction of attacking infantry. The crew were safe from the shrapnel inside the tank, but the enemy soldiers would be exposed. The Nahverteidgungswaffe could also be used to fire smoke grenades and signal flares. It looked like a large flare pistol. The commander's cupola was changed to a domed cast version.

In August 1943, the road wheels were strengthened with 24 outer rim bolts, but road wheels with 16 rim bolts were still being fitted to some Panthers as late as March 1944. Not all Panzer V Ausf.A Panther tanks looked the same.

Specifications

Length:	8.86m
Width:	3.42m
Height:	3.10m
Weight:	45.5 tonnes
Engine:	Maybach HL 230 P30 V-12 water-cooled 700hp petrol/gasoline
Crew:	5
Main gun:	7.5cm Kw.K.42 L/70
Other weapons:	2 x 7.92mm M.G.34 machine guns
Armour:	16 mm–80mm (Turret front 100mm–110mm)
Max. road speed:	55km/h
Max. range on roads:	200km
Total built:	2,200

Panther Ausf.A (Sd.Kfz.171)

Chapter 36
Panther Ausf.G (Sd.Kfz.171)

The Panzer V Panther tank was given the Ausf.G version designation to indicate this production run of tanks used a different redesigned hull. The turret and 7.5cm Kw.K.42 L/70 gun was the same one used on the earlier Ausf.A.

The side pannier armour that covered the top of the tracks on both sides of the tank was angled at 40 degrees on the Ausf.D and Ausf.A tank hulls. The new hull pannier side armour was sloped at 29 degrees. The thickness in the armour was increased from 40mm to 50mm. This increased the weight of the tank by 305kg.

To compensate for this increase in weight, the designers looked for areas where the thickness of the armour could be reduced. They chose to use 50mm armour plate on the lower front hull instead of the standard 60mm. This saved 150kg. The forward belly plates were reduced to 25mm from 30mm. The front two belly plates were 25mm thick, and the rear plate was 16mm thick. This saved a further 100kg in weight. The rear side armour wedges at the end of the superstructure were not part of the new design. The floor of the pannier was now a straight line. These weight reduction changes meant that the increase in side armour thickness did not result in an increase in weight of the Ausf.G tank hull compared with the older hull.

A perceived weak spot was the driver's armoured vision port, which was cut into the front glacis plate. This was removed in the design of the Ausf.G hull. The driver was provided with a single pivoting traversable periscope that was mounted in the roof of the hull and covered by an armoured rain shield. There were many other minor changes, including a new gun mantlet with a chin, which was fitted to some Ausf.G tanks, but the overall thinking behind the design was to simplify the construction process to enable more tanks to be built as fast as possible with some increase in crew protection.

Specifications

Length:	8.86m
Width:	3.42m
Height:	3.10m
Weight:	45.5 tonnes
Engine:	Maybach HL 230 P30 V-12 water-cooled 700hp petrol/gasoline
Crew:	5
Main gun:	7.5cm Kw.K.42 L/70
Other weapons:	2 x 7.92mm M.G.34 machine guns
Armour:	16 mm–80mm (turret front 100mm–110mm)
Max. road speed:	46km/h
Max. range on roads:	200km
Total built:	2,961 approx.

Panther Ausf.G (Sd.Kfz.171)

Chapter 37
Tiger I (Sd.Kfz.181)

Henschel was awarded the contract to build the Panzerkampfwagen VI Ausf.E (Sd.Kfz.181) Tiger tank. The first production Tiger was completed in August 1942. It and another seven produced in August were sent to the Eastern Front, south of Leningrad (St Petersburg). They first saw action on 16 September 1942 and the last of the 1,346 Tiger I tanks rolled out of the factory door in August 1944.

It had 100mm-thick, near-vertical, armour on the front hull and 100mm-thick armour on the lower front armour plate, angled at 25 degrees. The upper glacis plate was only 60mm thick but angled at 80 degrees. The gun mantle and the front turret armour gave a combined thickness of 145mm. In 1942, no Allied tank could penetrate the frontal armour of the Tiger tank.

When it first arrived on the battlefield, its accurate high-velocity long-range 8.8cm Kw.K.36 L/56 main gun could knock out all allied tanks at a distance of over 1km. This is why the Tiger tank built up a feared reputation. To help stop this 57-ton tank sinking into soft ground, it had very wide tracks and three interleaved road wheels on each suspension torsion bar to spread the load.

Early-production Tiger tanks had a drum-shaped cupola, rubber-rimmed road wheels and smoke grenade launchers on the side of the turret. Late-production Tigers had a domed-shaped cupola, metal-rimmed road wheels and side track-guards, and the smoke grenade launchers on the side of the turret were not fitted.

The Tiger I tanks saw action in the deserts of North Africa, Sicily, Italy, on the Eastern Front and in Normandy, following the 6 June 1944 D-Day landings. In December 1944, they took part in the Ardennes offensive, the Battle of the Bulge, and then the defence of Germany. Tiger I tanks were still in service at the end of the war, including during the Battle for Berlin.

Specifications

Length:	8.45m
Width:	3.70m
Height:	3.00m
Weight:	57 tonnes
Engine:	Maybach HL 210 P45 water-cooled 650hp petrol/gasoline (first 250 tanks) or Maybach HL 230 P45 V-12 water-cooled 700hp petrol/gasoline
Crew:	5
Main Gun:	8.8cm Kw.K.36 L/56
Other weapons:	2 x 7.92mm M.G.34
Armour:	15mm–100mm
Max. road speed:	45km/h
Max. range on roads:	125km
Total built:	1,346

Tiger I (Sd.Kfz.181)

Tiger I (Sd.Kfz.181)

Tiger I (Sd.Kfz.181)

German Tanks of World War Two

Tiger I (Sd.Kfz.181)

Chapter 38
Tiger II (Sd.Kfz.182)

The Panzerkampfwagen VI Ausf.B (Sd.Kfz.182) Königstiger (King Tiger) has more design similarities to the Panther tank than the Tiger I tank. It has thick, sloped frontal armour, making it virtually impregnable from the front. It was armed with a powerful and accurate high-velocity 8.8cm L/71 gun that could knock out any enemy tank. Königstiger also means 'Bengal Tiger' in German.

Henschel was awarded the contract to build 1,234 tanks. Only 492 were built by the time the Allies captured the factory in Kassel in March 1945. The front glacis armoured plate was 150mm thick, but the 50° angle meant that an enemy shell had to penetrate the equivalent of 230mm of armour. The sloped lower glacis plate was 100mm thick. The sides and rear armour of the tank were also sloped and 80mm thick. This thick armour meant that the tank was heavy. Therefore, to help dissipate some of the ground pressure, the King Tiger had very wide tracks.

Both the King Tiger tank turrets were designed and built by Krupp. The early-production curved Krupp turret is often wrongly called the Porsche turret. The curved front caused a 'shell trap' so only the first 50 tanks were fitted with this type of turret. The main-production Krupp turret with the flat front is also wrongly called the Henschel turret. The gun mantlet was 150mm thick, and the main production turret front armour was 180mm thick. The side and rear armour was 80mm thick.

The Königstigers first saw service in combat units in June 1944 and they were used on the Eastern and Western fronts. Allied troops encountered them for the first time in Normandy. They formed the spearhead of the Ardennes offensive, the Battle of the Bulge, in December 1944. They consumed a lot of fuel and caused re-supply problems. Some were abandoned because they ran out of fuel.

Specifications

Length:	10.28m
Width:	3.75m
Height:	3.09m
Weight:	69.8 tonnes
Engine:	Maybach HL 230 P30 V-12 water-cooled 750hp petrol/gasoline
Crew:	5
Main gun:	8.8cm Kw.K.43 L/71
Other weapons:	3 x 7.92mm M.G.42 (one anti-aircraft)
Armour:	40mm–180mm
Max. road speed:	35km/h
Max. range on roads:	140km
Total built:	489

Early-production, Krupp-designed, Tiger II turret.

Krupp-designed, Tiger II production turret.

Tiger II (Sd.Kfz.182)

Tiger II (Sd.Kfz.182)

The Swedish Königstiger, an early-production Tiger II used in heavy tank trials in Sweden.

Chapter 39
Panzer VIII Maus

Only two prototype Panzerkampfwagen VIII Maus (mouse) heavy tanks were built. They were known as vehicle VK 1000.01 typ-205-1 (V1) and vehicle typ-205-2 (V2) Maus.

The turret was to mount two guns side by side: a 7.5cm Kw.K.44 L/36 gun and the more powerful 12.8cm Kw.K.44 L/55 gun, the same gun that was fitted to the Jagdtiger.

The V1 Maus prototype received a mock weighted turret to enable cross-country field trials to proceed without having to wait for the finished turret. The V2 Maus prototype was fitted with the first completed turret.

The tank had extremely strong armour. The vertical hull sides were 180mm thick. The upper rear armour was 150mm thick and angled at 37°, and the lower 150mm plate was angled at 30°. The rounded front part of the turret was 220mm thick. The turret side armour was 200mm thick and angled at 30°. The turret rear armour was 200mm thick and angled at 15°.

The Maus was powered by a Daimler-Benz MB-517 V-12 water-cooled 44.5l 1,200hp petrol/gasoline engine that gave the tank a maximum speed road speed of 20km/h. The average road speed was 18km/h. The tracks were very wide (1.10m) to help decrease ground pressure.

Crossing most bridges in Europe was impossible. They could not take the 188 tonne weight. The Germans river crossing tactics was to have the tanks operating in pairs, the first crossing the river using electrical power, provided by a cable from the second, the air being supplied through a long snorkel.

As Soviet forces neared the tank-proving grounds at Kummersdorf, near Böblingen, the Germans decided to blow the two prototypes up so they did not fall into enemy hands. The hull on the V2 prototype was destroyed, but the hull of the V1 remained mainly intact as the explosives failed to explode properly. The Soviets placed the turret from the V2 prototype on the V1 hull to conduct tests after the war and this is on display in the Patriot Park Museum in Kubinka, Russia. The Maus was not used in combat.

Specifications

Length:	10.08m
Width:	3.70m
Height:	3.64m
Weight:	188 tonnes
Engine:	Daimler-Benz MB-517 V-12 water-cooled 44.5l 1,200hp petrol/gasoline
Crew:	6
Main gun:	12.8cm Kw.K.44 L/55
Other weapons:	7.5cm Kw.K.44 L/36 + 7.92mm M.G.34
Armour:	40mm–220mm (220mm front glacis plate)
Max. road speed:	20km/h
Max. range on roads:	160km
Total built:	2

Conclusion

The Allies won World War Two because of a massive advantage in manufacturing production capacity and access to raw materials that the Germans could not match. 'Armchair generals' talk about who had the best tanks, but professionals discuss who had the best logistic supply line. Many claim that it took on average four or five Allied tanks to knock out a Tiger tank in 1944–45. What is often not mentioned is the fact that there were always more tanks at the Allied supply depots to replace those knocked out. This was a luxury the German Army did not have. Many of the changes introduced during the production run of a German tank were to try and reduce costs and the use of certain difficult-to-obtain raw materials or to simplify manufacturing processes so more tanks could be built quickly. German High Command was very aware of production capacity problems and restraints.

This book has just covered German tanks used in World War Two. Many of the early tanks soon became outclassed and were no longer considered effective for front-line combat because either their armour was too thin or their main gun was not powerful enough. The German Army recycled many of these vehicles and used their hulls to construct anti-tank or artillery self-propelled guns. Panzer III and IV tank hulls were used in the construction of infantry support Sturmgeschütz assault guns. The Jagdpanzer IV tank hunter and Sturmpanzer IV were constructed using Panzer IV tank hulls. The Panther hull design was changed to build the Jagdpanther, and altered Tiger tank hulls were used to manufacture the Ferdinand (Elefant), Sturmtiger and Jagdtiger.

The German Army used many captured vehicles. Some were altered and used as flame-throwers, anti-tank guns or artillery self-propelled guns. The Italian Army produced their own range of armoured vehicles. When the country surrendered on 8 September 1943, many of their vehicles were entered into service with the German Army, who still occupied two-thirds of Italy at that time. The armies of German allies, such as Hungary and Romania, were supplied with German tanks, but they also built a range of tanks. These armoured fighting vehicles and a lot more will be covered in future volumes of Key Publishing's Military Vehicles and Artillery Series.